Jules Bernard Luys

The Brain and it's Function

Jules Bernard Luys

The Brain and it's Function

ISBN/EAN: 9783744679480

Printed in Europe, USA, Canada, Australia, Japan

Cover: Foto ©berggeist007 / pixelio.de

More available books at **www.hansebooks.com**

THE BRAIN

AND ITS FUNCTIONS

BY

J. LUYS

PHYSICIAN TO THE HOSPICE DE LA SALPÊTRIÈRE

WITH ILLUSTRATIONS

LONDON
KEGAN PAUL, TRENCH, & CO.,
1, PATERNOSTER SQUARE
1881

[The rights of translation and of reproduction are reserved.]

AUTHOR'S PREFACE.

THE present work, on the structure and functions of the brain, is an abstract of my personal experience as regards this subject, and of most of the ideas I have for many years been endeavouring to popularize in my public lectures at the asylum of La Salpêtrière.

It is divided into two very distinct parts.

The first, anatomical, serves as the foundation of the work. It is followed by a second, purely physiological, which is its complement and necessary sequence.

In the first part I have explained all the technical processes employed in arriving at the results indicated; insisting at the same time upon the value of the method which I have found it necessary to adopt, which consists in making regularly stratified sections of the cerebral tissue, in the faithful reproduction of these by means of photography, and in the employment of successively graduated powers for the representation of certain details.

I have been able, by means of these new methods of investigation, to penetrate further into the still unexplored regions of the nervous centres, and, like a traveller returned from distant lands, to bring back correct views and faithful representations of certain territories of which our predecessors caught scarcely a glimpse.

Thus, in fact, by making this photo-microscopic analysis of the nervous elements, I have been able to throw fresh light upon the intimate structure of the nerve-cell, and on the organization of its protoplasm, and to study it *in situ*, in its connections with the nerve-fibres and the surrounding network of neuroglia.

In my explanation of the grouping of the various portions of the cerebral mechanism, I have endeavoured as much as possible to simplify their description, and above all, to avoid employing that strange vocabulary now-a-days so improperly imported into the nomenclature of the different central regions of the brain.

I have therefore sketched synthetically the general economy of the structure of the brain, pointing out the intimate relations which exist between the cerebral cortex, the true sphere of psycho-intellectual activity, and the central ganglions (those of the optic thalami and corpora striata) which are in a manner the intermediate regions interposed between this and the excitations which proceed from the external world. I have insisted on the fact, which ten years ago I was the first in France to bring to light, namely, that the optic thalamus, with the isolated grey ganglions of which it is composed, represents a place of passage and reinforcement for excitations radiated from the sensorial periphery, while the corpus striatum, with its different compartments, and arches one within another, is on the contrary directly related to the passage of voluntary-motor excitations.

In this anatomical part I have particularly emphasized those details of the essential structure of the cerebral cortex, to the existence of which sufficient attention has not as yet been paid, and have utilized them from the stand-point of physiological interpretation.

Thus, having established the presence in the cerebral cortex of special zones of small cells subjacent to the pia-mater, and quite different in configuration from the zones of large cells occupying the deeper regions, I was led to see in this anatomical arrangement a clear relationship to a similar disposition existing in the constitution of the grey axis of the spinal cord.

As a consequence, I was led to think that if, as is experimentally demonstrated, the small elements in the spinal cord be affected by the phenomena of sensibility, it was natural to admit *physiological* analogies where *morphological* analogies exist; and consequently to consider the sub-meningeal regions of the cerebral cortex as being the histological territory specially reserved for the dissemination of sensible impressions; while the deeper zones of large cells (analogous to the anterior motor columns of the cord) might be considered as the regions of emission (psycho-motor centres) for exciting voluntary motion. Thus, I arrived at the demonstration that, in the very structure of the cerebral cortex, among the thousands of elements of which it is composed, there is an entire series of special nerve cells, intimately connected one with another, constituting perfectly defined zones, anatomically appreciable, and serving as a common reservoir for all the diffuse sensibilities of the organism, which, as they are successively absorbed in these tissues, produce in this region of the *sensorium commune* that series of impressions which brings with it movement and life.

In the second part, which comprises an explanati of the uses of the different cerebral apparatuses which the anatomical details have been previous analyzed, I have in the first place given a physiologi explanation of the different fundamental properties the nervous elements, considered as living histologi units.

I have in this manner shewn that these properti which are the ultimate generating elements of the forms of activity of cerebral life, may be fina reduced to three principal forms :—sensibility, by virt of which the cerebral cell enters into relation with t surrounding medium ; organic phosphorescence, wh confers upon it the property of storing up in itself a retaining the sensorial vibrations which have previou excited it (as we see in the inorganic world phospl rescent bodies preserve for a longer or shorter peri traces of the luminous vibrations which have imping upon them); automatism, which is merely the aptitu which the nerve-cell possesses, for reacting in preser of the surrounding medium, when once it has be impressed by this.

Having thus surveyed each of these elementary p perties of the nervous elements in their origin, in th evolution throughout the organism, in their normal ma festation and pathological deviation, I arrive at demonstration that it is by means of their combinati and· by the harmonious co-ordination of all their tr specific energies, that the brain feels, remembers, a reacts ; and that, in fact, being the properties in which the others originate, they are the only living forces t are always present, always underlying the infinite ser of operations which it every moment accomplishes; and that without them that admirable and complex apparatus, at once so delicate and so simple, would be as absolutely without life and without movement, as the earth would be without the sun.

Having thus examined the elementary properties of the nervous elements, I have shewn how their co-operation may be used to explain the principal phenomena of cerebral physiology.

I have in this manner made it clear that by grouping among themselves the foregoing data, we may perceive that all manifestations of cerebral activity— even though we have to deal with the phenomena of psychical life proper, or the operations of intellectual life,—like their fellows which have the spinal cord for a theatre (reflex phenomena) are always susceptible of being decomposed into three elementary phases; that they are always originally determined by the arrival of an incident sensorial impression, recent or former (phase of incidence); accelerated by the particular reaction of the interposed medium, reacting by virtue of its specific energy (intermediate phase); and completed by the secondary reaction of the intermediate medium, reacting and carrying outwards the primordial vibration which has been communicated to it (phase of reflexion).

It results, then, from this manner of looking at the phenomena of cerebral activity, that it is always a fact of the vital order which is at the origin of every process in evolution. Sensibility is always the primary motor agent; it originates all movement. Propagated through the sensori-motor machinery of the cortex, it becomes insensibly transformed, like a force in evolution, and ends by disengaging itself from the organism in the form of a motor act.

In short, in these researches, in which my sole object has been to carry the data of contemporary physiology into the hitherto uninvaded domain of speculative psychology, I have endeavoured to show that the most complex acts of psycho-intellectual activity are all definitely resolvable, by the analysis of nervous activity, into regular processes; that they obey regular laws of evolution; that, like all their organic fellows, they are capable of being interrupted or disturbed in their manifestations by dislocations occurring in the essential structure of the organic substratum which supports them; and that, in a word, there is from this time forth a true physiology of the brain, as legitimately established, as legitimately constituted, as that of the heart, lungs, or muscular system.

As a consequence of what has just been said, it necessarily follows that this range of studies, so new and so attractive, should properly belong to the physiological physician and to him alone. Henceforward he may claim as his peculiar patrimony that special domain of the nature of man concerning which speculative philosophy has for so many centuries so long and learnedly harangued. It will be his task to fertilize it by his incessant labour, and to make it yield what all labour intelligently directed should afford, legitimate fruits—practical consequences which may be utilized for the benefit of suffering humanity. The history of medical science is present with its daily lessons, to shew us that the useful acquisitions which it has made have always been inevitably subordinated to clearer and more precise notions concerning the anatomy of the organs whose care is its mission; and when we transfer the same aspirations to the subject which now occupies us, this fact surely authorizes us to hope that in the future we shall see new methods in the treatment of mental maladies, and modes of action more efficacious than those now at our disposal, arise from a better comprehended cerebral anatomy, and a more rationally directed cerebral physiology.

J. LUYS.

CONTENTS.

PREFACE . . . . PAGE v

PART I.
ANATOMY OF THE BRAIN.

CHAPTER I.
METHODS OF STUDY 1

CHAPTER II.
CORTEX OF THE BRAIN—THE GREY CORTICAL SUBSTANCE . . 11

CHAPTER III.
THE WHITE SUBSTANCE OF THE BRAIN 26

CHAPTER IV.
THE OPTIC THALAMUS 34

CHAPTER V.
THE CORPUS STRIATUM 46

CHAPTER VI.
PHYSIOLOGICAL DEDUCTIONS 59

CHAPTER VII.
Physico-Chemical Phenomena of Cerebral Activity . .

PART II.

GENERAL PROPERTIES OF THE NERVOUS ELEMENTS.

BOOK I.

SENSIBILITY OF THE NERVOUS ELEMENTS.

CHAPTER I.
Graduation and Genealogy of the Phenomena of Sensibility . . . .

CHAPTER II.
Evolution of the Process of Sensibility through the Mechanism of the Nervous System—Unconscious Sensibility—Conscious Sensibility (Sensation) . . . .

CHAPTER III.
Intra-Cerebral Propagation of the Processes of Sensibility

CHAPTER IV.
Perturbations of Sensibility . . . .

CHAPTER V.
Development of Sensibility . . . .

BOOK II.

ORGANIC PHOSPHORESCENCE OF THE NERVOUS ELEMENTS.

CHAPTER I.

	PAGE
INTRODUCTORY	133

CHAPTER II.

GENESIS AND EVOLUTION OF MEMORY 142

CHAPTER III.

THE MEMORY IN EXERCISE. 150

CHAPTER IV.

DEVELOPMENT OF THE PHENOMENA OF MEMORY 159

CHAPTER V.

FUNCTIONAL DISTURBANCES OF THE PHENOMENA OF MEMORY . 165

BOOK III.

AUTOMATIC ACTIVITY OF THE NERVOUS ELEMENTS.

CHAPTER I.

INTRODUCTORY 171

CONTENTS.

CHAPTER II.
GENESIS AND EVOLUTION OF AUTOMATIC ACTIVITY . . . . PAGE 176

CHAPTER III.
AUTOMATISM IN PSYCHO-INTELLECTUAL ACTIVITY 180

CHAPTER IV.
DREAMS 195

CHAPTER V.
DEVELOPMENT OF AUTOMATIC ACTIVITY 200

CHAPTER VI.
FUNCTIONAL PERTURBATIONS OF AUTOMATIC ACTIVITY . . . 205

PART III.
EVOLUTION OF THE PROCESSES OF CEREBRAL ACTIVITY.

BOOK I.
PHASE OF INCIDENCE OF THE PROCESSES OF CEREBRAL ACTIVITY.

CHAPTER I.
ATTENTION 215

CONTENTS. xvii

CHAPTER II.
CONSTITUTION OF THE SPHERE OF PSYCHO-INTELLECTUAL ACTIVITY 226

CHAPTER III.
GENESIS OF THE NOTION OF PERSONALITY 233

CHAPTER IV.
DEVELOPMENT OF THE NOTION OF PERSONALITY 238

CHAPTER V.
FUNCTIONAL DISTURBANCE OF THE NOTION OF PERSONALITY . 244

BOOK II.
PHASE OF PROPAGATION OF THE PROCESSES OF CEREBRAL ACTIVITY.

CHAPTER I.
DISSEMINATION OF SENSORIAL IMPRESSIONS IN THE PLEXUSES OF THE PSYCHO-INTELLECTUAL SPHERE—GENESIS OF IDEAS . . 250

CHAPTER II.
EVOLUTION AND TRANSFORMATION OF SENSORIAL IMPRESSIONS . 256

CHAPTER III.
THE JUDGMENT 289

BOOK III.

PHASE OF REFLEXION OR EMISSION OF THE PROCESSES
OF CEREBRAL ACTIVITY.

CHAPTER I.

REFLEXION OF MOTOR PROCESSES UPON THE PHENOMENA OF VEGE-
TATIVE LIFE 313

CHAPTER II.

TRUE PERIOD OF EMISSION OF THE PROCESSES WHICH PRODUCE
VOLUNTARY MOTION — SPONTANEOUS REACTION OF THE
SENSORIUM—MOTIVED RESOLUTION 319

LIST OF ILLUSTRATIONS.

	PAGE
FIG. 1.—DIAGRAM OF A SECTION OF THE CEREBRAL CORTEX	15
FIG. 2.—CORTICAL CELL OF THE DEEPER ZONES	19
FIG. 3.—DIAGRAM OF COMMISSURAL FIBRES OF THE ANTERIOR REGIONS OF THE BRAIN	27
FIG. 4.—DIAGRAM OF COMMISSURAL FIBRES ON THE LEVEL OF THE CORPUS STRIATUM	29
FIG. 5.—DIAGRAM OF CONVERGING FIBRES, AND THEIR RELATIONS WITH THE CENTRAL GREY GANGLIONS	31
FIG. 6.—DIAGRAM OF THE SENSORI-MOTOR PROCESSES OF CEREBRAL ACTIVITY	61

THE BRAIN.

PART I.

ANATOMY OF THE BRAIN.

CHAPTER I.

METHODS OF STUDY.

THE study of the nervous centres has always strongly attracted the anatomist as a field of labour; and the reason of this is not far to seek. In the face of such a subject, not only does the very natural desire to penetrate the inmost secrets of the organization of the anatomical details under consideration come into play, but, further, there is that unconscious attraction which draws the human mind towards the unexplored regions of the unknown—towards those mysterious realms where the living forces of all our mental activities are silently elaborated, and where the solution of those eternal problems, regarding the relations of the physical organization of the living being to the acts of its psychic and intellectual life, evades us as we pursue it.

Hence it is that from century to century most of the great anatomists have, each in his turn, laboured in this direction. Hence Galen, Varolius, Willis, Malpighi, Vieussens, Vicq d'Azyr, Sömmering, Reil, etc., have successively, in their immortal works, either described the organization of the nervous centres as they conceived of it at their own epoch, or expressed in their iconographies (with a more or less distinct glimpse of the truth) the objective fashion in which they saw the anatomical details they have successively represented.

In dealing with a subject so vast and so delicate, and a material so fragile and easily alterable as the nervous matter, the student is necessarily forced to depend on the different methods placed at his disposal by the arts and sciences of his own epoch. Hence the smallest technical discoveries frequently become of inestimable value; and it may be said, without exaggeration, that the utilization of chromic acid,* which, by hardening the nerve-substance, fixes it, with all its natural relations, without altering it, has been one of those new methods in laboratory work which have most essentially contributed to the success of those great achievements in this particular domain of anatomical science which our own century has witnessed.

On the other hand, the perfecting of the magnifying power of microscopes has been of immense service, and has permitted the spirit of man to advance with vast strides into regions as yet unexplored, where it stands face to face with those ultimate anatomical units, the nerve-cells, of which our predecessors scarcely

* Hannover, in 1840, was the first to point out the hardening properties of chromic acid. (Robin, Traité du microscope, p. 297. J. B. Baillière, 1871.)

caught a glimpse. Thus it is now possible to give exact descriptions of their configuration, whether we study their connections, their minute structure, or the different pathological deviations they may undergo.

The introduction of the microscope into the study of histology has been in our century for the world of the infinitely little, what at another period of human development the intervention of the telescope was for the exploration of the sidereal world. It has rendered distinctly visible all those myriads of elements which, from their extreme smallness, were concealed from the eyes of our predecessors. It has brought them to light, revealed the secrets of their minute organization, and opened to the investigations of anatomists an entire new world of unexpected ideas.

Following upon this discovery, as a natural consequence, came the revelation of the art—previously unknown in our laboratories—of making thin slices of nervous tissue, colouring them, rendering them transparent, and preserving them. The employment of reagents of all kinds, which, testing in some degree the special sensibility of each histological element, colours it in a particular manner, and sets in relief the peculiarities of its structure, has opened a new road for progress; so that all over the civilized world, labourers, aided by physics and chemistry, have united their efforts, until we can say that the limits of the unknown recede, and that new conquests are perpetually being registered in our scientific reports.

But this is not all. In this kind of research it is not sufficient to see for ourselves the new facts met with on our road; it is necessary to make others see them, to represent in faithful statements the details of nature we have examined, and to place the newly-registered facts beyond dispute.

Up to the present time it was the observer himself who pourtrayed, by means of his pencil, the objects which passed through the focus of his microscope. And, accordingly, we all know how widely these nominal drawings—even those made by masters of their profession—usually diverge from the truth; simply because they can never express more than those details which the artist has perceived and recognized, and a species of unconscious selection from the objects which are passing before his eyes. It is, then, in presence of these desiderata, as regards graphic representation, in drawings made by hand that we feel the necessity of applying the marvellous resources now offered us by photography to the reproduction of microscopic objects.

The sensitized plate henceforward plays its part in the world of scientific investigation, in the study of the phenomena that occur in the world of the infinitely little, as well as in the study of those that occur in the world of the infinitely great—registering histological facts as well as astronomical phenomena, and thus becoming the impersonal and automatic pourtrayer of the most minute details that have impressed themselves upon it. Thus, wonderful to relate, photography, very much superior to drawing, not only reveals the objects which the eye perceives, but brings to light in addition a whole series of latent details, which await but the intervention of a simple lens to be successively recognized upon the prints when obtained.

These new means of investigation, which the scien-

METHODS OF STUDY.

tific methods of the nineteenth century have placed within the reach of our generation, will, therefore, explain the progress accomplished, and show us once more that, in the long process of evolution which extends through ages, man only arrives step by step at the fragments of truth which he snatches, and that even his most persevering efforts only serve to cause the unknown to recede a few paces backwards. It is strange to find that, as fast as any progress is accomplished and new discoveries registered, new problems incessantly start up; and that just when we thought we had arrived at the utmost limits of the known world, at the demonstration of elements, simple, fixed, definite, our perfected methods of study enable us to see new complexities and unexpected horizons.

Thus, for instance, by means of high powers, the histological elements of the nerve-cell, hitherto considered as the primordial and irreducible units of the system, become themselves divisible into secondary elements.

Photo-chemical histology, indeed, shows us that the protoplasm of the cell, formerly described as a homogeneous substance, is arranged in a fibrillary trellis-work; that its nucleus presents an arrangement of radiated fibres; and that what was thought to be the nucleolus is itself a complex element. The nerve-cell thus becomes in its turn a little nervous organ *sui generis*. (*See* Fig. 2.)

The same analytic processes enable us, moreover, to demonstrate that the network, so dense and compact, which unites all the nerve-cells of the cerebral cortex, for instance, one with another, is so delicate that, when enlarged to 286 diameters, the fibres of which it is composed become visible, like single hairs in appearance and magnitude, etc.

What will be the end of these unforeseen details which present themselves in the train of each adaptation of a new method of study, to our researches into the nervous system?

No one knows as yet. It seems as though the secrets of nervous organization escape from our eyes as fast as we press further into the regions where they conceal themselves, and while anticipating the new methods of analysis which the future holds in reserve, we cannot help thinking that there is still much to do, and that now, more than ever, we should remember that true saying of Serres: "We have been dissecting the brain since Galen's time, yet there is not an anatomist who has not left his successors something to do."

The labours of which I am about to give a *résumé*, are, then, but one of the phases of this long discussion concerning the structure of the nervous centres which has been going on for centuries.

If they do not establish the truth absolutely and finally, they will at least have the merit of being the result of contemporary science, and a sort of synthesis of the methods of work at our disposal.

The method I have employed for studying the organization of the cerebro-spinal centre in man, I have already explained in my first work.* It essentially consists in the preparation of a series of sections made methodically, millimetre by millimetre, vertically, hori-

* J. Luys, "Recherches sur l'anatomie, la physiologie et la pathologie du système nerveux." Paris, 1865, J. B. Baillière.

zontally, and antero-posteriorly; and—these sections being thus made according to the three dimensions of the solid mass which was to be studied—in reproducing them all photographically.

I set myself, then, to make a series of successive horizontal sections of the brain, previously hardened in a chromic acid solution, from apex to base, at intervals of about one millimetre, and as perfect as possible; each being in its turn reproduced by photography.

I made similar sections of the brain in a vertical and antero-posterior direction, and at regular intervals from behind forwards.

These operations having been thus regularly conducted, this method enabled me to have representations of the reality as exact as possible; to keep the natural relations of the most delicate portions of the nervous centres each by each, according to their normal connections, and, in fact, without deranging anything. Thus by comparing the sections, horizontal or vertical, one with another, I could follow a given order of nerve-fibres in its progress, see its point of origin, and its point of termination; study the natural increase in complexity of the different kinds of nerve fibrils, millimetre by millimetre, changing nothing, lacerating nothing, and leaving everything pretty much in its normal position.*

* The plan of this work does not permit me to insist upon the innumerable difficulties I have surmounted, in arriving at the clear result already recorded in my photographic iconography. (Luys, "Iconographie des centres nerveux," J. B. Baillière, Paris, 1872.)

In the first place I had to invent cutting instruments sufficiently delicate to make complete sections of the brain, of the thickness of about one millimetre.

But these pieces, when sufficiently hardened to undergo the action of the cutting instrument, had acquired, on coming out of the bath of chromic acid,

By means of these new photographic methods of reproduction, which are all the more precise because impersonal, I had only, then, to register the details the sun himself had printed, to place the prints in juxtaposition, to compare them one with another, and thus to make a single synthesis of the multiple elements of analysis I had thus obtained by the automatic co-operation of the light.

The general view of cerebral topography having thus been fixed by these processes, the regions of more delicate texture, the special points which it was necessary to study in their minute elements, were further sufficiently magnified and reproduced, with successively increasing powers. I could thus render visible to the naked eye, and exhibit on a plan, details of structure which, up to that time, had only been seen in isolation under the tube of the microscope. By this means the mind of the observer, penetrating successively from the known to the unknown, from well-defined regions to those which are not so as yet, can easily make itself familiar with the details of the minute structure of the final nerve elements.

The cerebro-spinal system in man and the vertebrates consists of three departments, independent one of another, and yet very intimately connected. These are:—

1. The *cerebrum* proper.

that peculiar uniform greenish colouring which renders them completely unfit for photogenic action. It was therefore necessary to discover a perfectly novel series of processes, in order to purify these sections from the chromic acid, and, without altering them, to impart to them photogenic properties. (See *Journal d'Anatomie de Robin*, Paris, 1872, for the whole series of the processes employed to bleach the sections tinted with chromic acid.)

METHODS OF STUDY.

2. The *cerebellum* and the apparatuses of cerebellar innervation annexed thereto.

3. The *medulla spinalis* and its encephalic expansions.

In this study we shall occupy ourselves with the cerebrum proper only.

The cerebrum consists of two lobes or hemispheres united to one another by a series of white transverse fibres, which form an anastomosis between the homologous regions of each lobe, so as to constitute a twin system, of which all the molecules are consonant one with another.

Each cerebral lobe, taken alone, presents for consideration in its turn :—

1. Masses of grey matter.
2. Agglomerations of white fibres.

The masses of grey matter, which are composed of many myriads of cells, and are the essentially active regions of the system, are arranged at the periphery in the form of a thin, undulating, continuous layer, which constitutes the cerebral cortex; and in the central regions in the form of two grey ganglions, coupled together, which are simply the grey substance of the optic thalami and corpora striata (opto-striate ganglions).

The white substance, essentially composed of nerve-tubules in juxtaposition, occupies the spaces comprised between the cortical periphery and the central ganglions.

The fibres of which it consists, and which merely represent lines of union between such and such regions of the cortical periphery and such and such regions of the central ganglions, run, like a series of electric wires stretched between two stations, in two principal directions.

1. Some directly unite the different points of the cortical periphery with the central ganglions, and are lost in their mass.

These are like the spokes of a wheel which unite its circumference to the central nave, which serves as their point of support. We may therefore describe them under the name of *converging fibres.*

2. The others, on the contrary, have a transverse direction. They proceed from one hemisphere to the other, thus uniting the homologous regions of the brain, right and left.

It may therefore be said that they serve as an anastamosis and commissure between these homologous regions, and that they are thus the agents which produce unity of action between the two cerebral hemispheres. This order of fibres, by reason of its origin and connections, may legitimately be designated by the name of *commissural fibres.*

These data being admitted, it may be said that the anatomic formula by means of which we may define the structure of the cerebrum, of man as of the other vertebrates, is this: "The cerebrum is the sum total of the cerebral convolutions, united one with another, with those on the same side and with those on the other, and simultaneously with the central opto-striate ganglions."

We shall now pass in review the different agglomerations of the grey matter, and at the same time give a sketch of the principal details of the organization of the white matter.

CHAPTER II.

CORTEX OF THE BRAIN—THE GREY CORTICAL SUBSTANCE.

EVERY one knows the external appearance of the cortical substance of the brain. It is sufficient to recall that of the brains of sheep, as served at table, to see at a glance that the grey cortical substance presents the appearance of a grey undulating layer, folded a great number of times upon itself, and thus forming a series of multiple sinuosities of which the sole object is the obtaining of increased surface.

These foldings and refoldings, which attain their maximum of development in the human species, apparently obey some fixed laws as regards their distribution.* Some, in fact, have permanent characters which render them easily discoverable in all human brains; others, and these form the greater number, present all possible varieties of external configuration, not only in different individuals, but even in the same individual, according as we inspect homologous regions in the right or left hemisphere.

Take, for instance, a sheet of tracing paper, apply

---

* See the interesting description of the topography of the cerebral convolutions given by Prof. Charcot in his lectures to the Faculty.—Progrès Médical, 1875, p. 283, 353, &c.

it to a fresh vertical section of the brain, mark with a brushful of water-colour the contour of the cortical substance of one hemisphere, and fold the paper over; you will thus see very clearly that the outline of the convolutions of one side does not adapt itself to those of the other. I have made such tracings repeatedly, and have never yet found a human brain completely symmetrical, completely balanced in its peripheral regions, and with the left regions of the cortical substance exactly corresponding to the homologous regions of the opposite side.

There is another peculiarity, which it is important to notice, in the external examination of the cortical substance.

In the adult, in vertical or horizontal sections of the brain, it is evident that the line of the summits of the convolutions is continuous, that their culminating points are all on the same level; there is some uniformity in the distribution of the activity of nutrition over the whole mass.

As old age advances different appearances begin to show themselves, and in studying the different effects of senescence in all the organs, it is curious to observe its characteristics in the human brain.

We observe, then, that the grey substance becomes diminished in thickness; that its colour changes to yellowish white in consequence of the passing of the nerve-cells into the granulo-fatty state; and that besides, the convolutions settle down in isolated groups, like mountains, undermined at their bases, which insensibly subside. Thus, in many old men in their dotage, we may note that the line joining the summits of certain groups of convolutions becomes interrupted ; that a certain number of them are retracted and have sunk below the level of the surrounding convolutions; and that thus, from the effect of time, there exists a slow and progressive absorption of the nervous substance.

In individuals who fall prematurely into dotage from alterations of the cerebral substance, under the action of mental diseases, we find the same atrophy of the cortical layer. Thus I have very frequently observed atrophy of the convolutions in young subjects attacked by paralytic dementia, persons affected by hallucinations, and patients who have suffered from melancholic delirium.

The thickness of the cortical substance in the adult is on the average about two to three millimetres. Generally it is more abundant in the anterior than the posterior regions. Its mass varies according to age, and especially according to race, Gratiolet remarking that in races of low stature the mass of the cortical substance is but small.*

Its colour presents some varieties. It is uniformly greyish, and as it were gelatinous, in the new-born infant; in the child during its first years it is of a rosy grey; in the old man it acquires somewhat of a yellowish-white colour, its vascularity being less distinct than in the adult. In the negro this substance is of a darker colour than in the white man.

In the adult in whom development is regularly accomplished, the cortical substance presents itself very clearly to the naked eye, in the form of stratified zones, differing slightly in colour. We observe, in fact, that

* "Gratiolet, Bulletin de la Société d'Anthropologie," 1859, p. 38.

there exists a superficial sub-meningeal zone of a greyish colour, and transparent; and a deeper zone, underlying the preceding, of a more distinctly reddish colour.

When we take a thin section of this cortical substance, compress it between two pieces of glass, and examine it by holding it up to the light, as Baillarger first pointed out,* we see that it divides into secondary zones of unequal transparency, and that these zones cleave with a regular and fixed striation. We shall see that these appearances are merely the result of the minute structure of the cortical substance.

Such are the characters which the cerebral cortex presents to the naked eye, and which every one may observe in fresh brains.

Let us now penetrate, by means of magnifying glasses, into the interior of this soft substance, amorphous in appearance, of which the homogeneous aspect is far from revealing to us its marvellous details.

Let us push our researches still further by means of thin sections rendered transparent and methodically coloured; let us employ gradually increasing powers to pass from a known to an unknown region; and avail ourselves of the magnifying processes that photography places at our disposal. We shall then be able to penetrate into these almost unknown regions of the world of the infinitely little, and, like travellers returned from distant lands, to bring back various photographic images—indisputably faithful reproductions of the details which have struck us in the course of our voyage of discovery.

We now find in the cortical substance a fixed

* "Mémoires de l'Académie de Médecine de Paris," 1840.

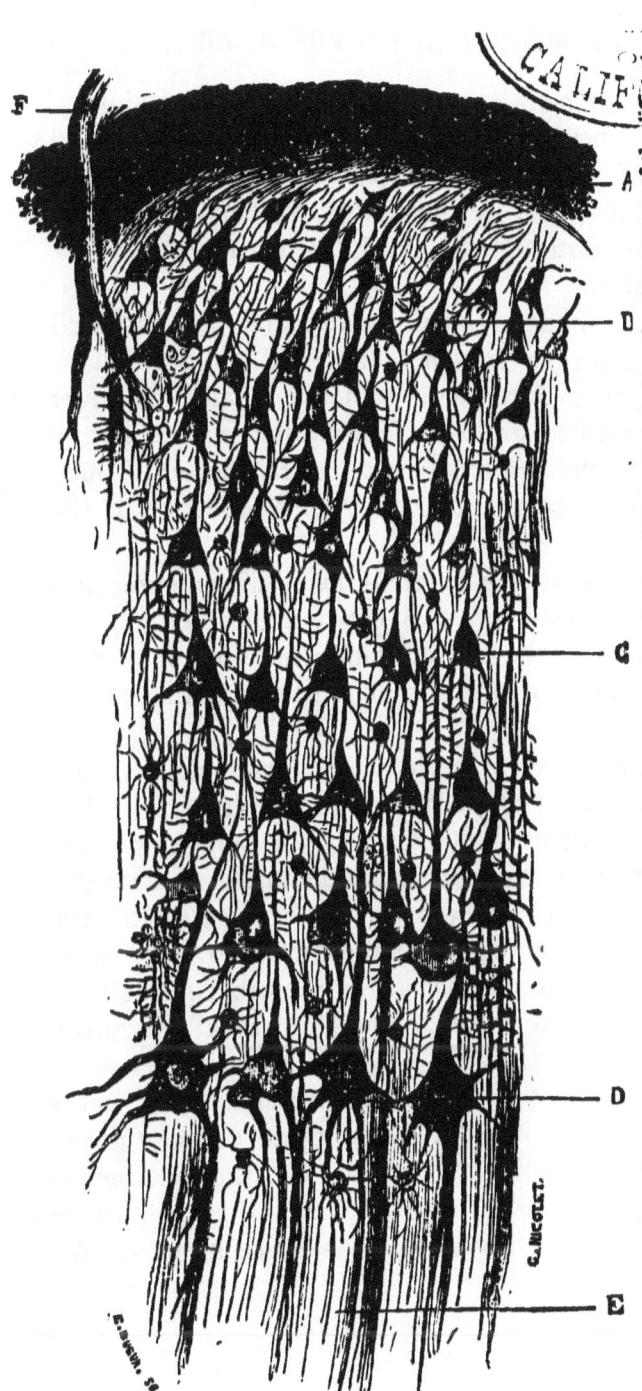

FIG. 1.—Half-diagrammatic figure of the cerebral cortex, magnified about 280 diameters, giving a view of the entire arrangement of the different zones of cells, and their relations to one another, and to the surrounding neuroglia. The region A corresponds to the sub-meningeal network of the neuroglia. The region B to the sub-meningeal zones of small cells (region of the *sensorium commune*); the region C is intermediate between the sub-meningeal and the deeper zones of cells which are indicated af D. At E we note the dipping of the fasciculi of white substance into the plexus of cortical cells. F represents a capillary at the moment when it plunges into the tissues of the cortex.

anatomical element—an ultimate morphological unit. This is the nerve-cell, with its various attributes and definite configuration, its nerve-fibres, connective-tissue, and capillaries; and we must now examine the constitution of this cortical nerve-cell, its forms, its connections, and its relations.

Imagine a number of small pyramidal bodies, disposed in series, parallel to one another, united to one another by means of an intermediary network, and moreover regularly stratified, and thus forming layers successively piled up, like the strata of the terrestrial cortex. Such is the general aspect that a thin complete section of the cortical substance presents.

If we add that the cerebral nerve-fibres enter into intimate connection with this network of cells, and are insensibly lost in the surrounding tissue, we shall then have a complete expression of the organization of the cerebral cortex.

Now if we observe each of the nerve-cells singly, we discover that they all have a pyramidal form; that they are of unequal volume; that the smaller occupy the superficial or sub-meningeal, the larger the deeper regions; that these latter are on an average double the size of their fellows, and that the transition from small to large cells is accomplished by insensible gradations, the cells of the intermediate zones in general presenting mixed characteristics.

The cells have in addition one extremely remarkable peculiarity, which gives to the histological preparations of this region a special physiognomy, viz., their characteristic arrangement. It is indeed very curious to observe that, while they are all, as we have seen, pyramidal

THE GREY CORTICAL SUBSTANCE. 17

in form, the summit of each is, so to speak, attracted towards the superficial regions, like a series of needles magnetized so as to point towards the pole; so that their bases are all parallel and are directed towards the point from which the nerve-fibres arrive.

They give off from their substance a species of very delicate, rootlet-like, hirsute fringe, which gradually spreads out and forms on all sides a surrounding network; and as each cell presents a similar arrangement, the result is that a close union between them is produced, so that they form throughout the cerebral cortex a continuous true plexus, all the molecules of which are by some means arranged so as to vibrate in unison.

By their prolongations, which form the base of the pyramid, they enter more or less directly into relation with the afferent nerve-fibres; while their apices send out a filamentous prolongation, which proceeds either to be lost in the surrounding network, or to enter into contact with certain zones of cells situated above.

The number of cells in the cortical substance must be estimated at many thousands. The following data are sufficient to make this clear.

In a space of cortical substance equal to 1 square millimetre, and of a thickness of $\frac{1}{10}$th of a millimetre, 100 to 120 nerve-cells of various sizes have, on an average, been counted. If now we form in imagination an estimate of the ratio of this small portion of the cortical substance to the whole, we shall arrive at an estimate of many thousands.

The colour of the cortical cells in fresh healthy brains

is an amber yellow. They are apparently provided with a bright nucleus and with a nucleolus.*

The internal structure of the cerebral cell individually considered, seems to grow more complex the deeper we proceed in the minute study of its elements.

Some years ago anatomists admitted of an investing membrane, and a contained substance, with a nucleus and a nucleolus, in the constitution of the cell; later they discovered that its investing membrane was nothing more than the external layer of an amorphous protoplasm, surrounding the nucleus of the cell, and prolonging itself externally in the form of multiple ramifications.

At the present day things seem to complicate themselves still further; for I have lately discovered in some researches as yet unpublished, that this substance which we call the protoplasm of the cell is formed by a true tissue organized in a special manner; that this tissue, consisting of very delicate fibrillæ interlaced like

---

* The cells of the cortical substance were perfectly described by Malpighi in the year 1687, and, strange to say, left in oblivion by the majority of anatomists during the interval between him and us. It is only in our time that they have been more definitely brought to light. "I have thus discovered," says Malpighi, "by the dissections I have made of the brains of sound animals, that the cortical substance of the brain consists of numbers of small glands piled up and united together. These glands, in which are inserted, or rather from which spring the white roots of the nerves, are so intricately arranged, joined, and connected with one another in these regions of the brain, which resemble little tangled thickets, that they form by their assemblage the cortex or external superficies of the brain. They are of an oval shape, which is, however, always more or less flattened, from their pressing one upon another on all sides. From their internal parts there springs a white nerve-fibre, which is as it were the vessel belonging to them, and which may be clearly seen through these small transparent and entirely white bodies: so that the white substance of the brain is apparently a tissue and an assemblage of several sorts of small fibres joined together, etc. etc." ("De la structure des viscères," Paris, 1687.)

THE GREY CORTICAL SUBSTANCE. 19

the wicker-work of an osier-basket, has a tendency to agglomerate towards the nucleus of the cell, which thus

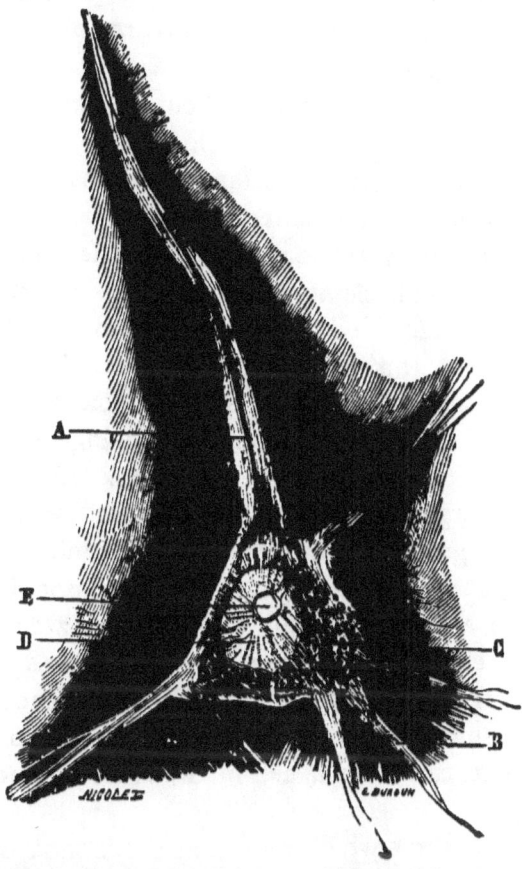

FIG. 2.—Cortical cell of the deeper zones at about 800 diameters; a section of the cell is made through its greater axis, its interior texture being thus laid bare. A represents the superior prolongation radiating from the mass of the nucleus itself; B, lateral and posterior prolongations; C, spongy areolar substance, into which the structure of the cell itself is resolved ; D, the nucleus itself seems only to be a thickening of this areolar stroma—it sometimes has a radiated arrangement; E, the bright nucleolus is itself decomposable into secondary filaments.

becomes a true point of concentration ; that the nucleus

itself is not homogeneous; that it is endowed with a special structure, radiated in appearance; and that lastly the nucleolus, considered as the final expression of the unity of the nerve-cell, is in its turn divisible into secondary filaments.

Imagination is confounded when we penetrate into this world of the infinitely little, where we find the same infinite divisions of matter that so vividly impress us in the study of the sidereal world; and when we thus behold the mysterious details of the organization of an anatomical element, which only reveal themselves when magnified from 700 to 800 diameters, and think that this same anatomical element repeats itself a thousandfold throughout the whole thickness of the cerebral cortex, we cannot help being seized with admiration; especially when we think that each of these little organs has its autonomy, its individuality, its minute organic sensibility; that it is united with its fellows; that it participates in the common life; and that above all it is a silent and indefatigable worker, discreetly elaborating those nervous forces of the psychic activity, which are incessantly expended in all directions and in the most varied manners, according to the different calls which are made upon it, and set it vibrating.

The nerve-fibres which represent the bonds of union between the cortical substance and the central regions of the brain emerge from the midst of the plexus of cells. They all at first appear as isolated filaments, as a derivation, mediate or immediate, from the tissue proper to each cell; then by degrees, as they proceed between the ranges of cells, they enlarge, their

THE GREY CORTICAL SUBSTANCE.

sheath thickens, the interposed fatty substance becomes more abundant, and they are insensibly transformed from grey to white fibrils. As to their mode of central arrangement they behave in a manner which we shall explain further on.

Neuroglia.—Among the elements which enter into the structure of the cerebral cortex, the uniting substance, the *neuroglia* as it is called, plays a primary part as regards its anatomical connections and physiological properties.

Imagine a web of extreme delicacy, radiating from the walls of the sheaths of the capillaries of the cerebral membranes, and immediately enveloping the cortical substance; its prolongations, like an infinite number of rootlets, everywhere plunging into this substance. Imagine this delicate web, having resolved itself into a network of greater and greater attenuation, forming meshes more and more closely woven, in the midst of which the nerve-cells are, to borrow the picturesque simile of Malpighi, embedded like pomegranate seeds, in the midst of the white fibrous tissue which encloses them on all sides.

These same neuroglian filaments thus envelop the nerve-cells with their inextricable web, just as the cellular tissue, for instance, surrounds the lymphatic ganglions; and thus is constituted that immense network of connective-tissue everywhere continuous throughout the nervous system, from the spinal-cord to the brain, serving to support all the individual anatomic elements, and by its softness, delicacy, and extreme divisibility, forming for them a veritable *cement* which solders them together, uniting them in a perfect unity, while at the same time it serves them as a means of nutrition.

This network of the neuroglia presents, moreover, a very remarkable arrangement in the cortical substance. Not only does it incorporate itself with each particular cellular element, and with the nerve-fibres, serving them in a manner as a mechanical protection, but besides this it plays an analogous part as regards the nervous elements of the cortex as a whole.

Thus, if we examine the superficial layers of each convolution in the sub-meningeal regions, we perceive that the neuroglia forms, immediately above the last zones of nerve-cells, a thin areolar layer of an appreciable thickness, constituting a sort of spongy cushion everywhere continuous. It is, in fact, a means of protection and isolation which, as it were, filters the nutrient juices flowing from the meninges, and prevents the plexus of nerve-cells, thus protected by this variety of natural epithelium, from coming nakedly into direct contact with the capillaries of the meningeal membranes. (See Fig. 1, A.)

The capillaries similarly play a very important part in the structure of the cortical layer. They represent the most important of the nutritive elements that bring to the nerve-cell the *pabulum vitæ* necessary for the maintenance of its daily activity.

Radiating in the form of little canals from the deep surface of the meninges, they plunge like very delicate rootlets into the midst of the nervous elements, dividing themselves into a network of greater and greater tenuity, and their meshes, becoming closer, pass around the circumference of each group of cells to form areolæ extremely rich in blood-vessels. It is a very remarkable fact that these same capillaries, which directly penetrate the texture of other organs and come in contact with the active elements which it is their task to nourish have a special arrangement as regards the nervous elements. A peculiar adventitious sheath, in fact, surrounds their walls, like a muff, for a part of their circumference, isolating them from the nervous elements themselves; so that it is but mediately that these obtain their share in the processes of nutritive life.

To sum up, the structure of the cerebral cortex may be reduced to the following propositions :—

The cortical substance is composed of fixed anatomical elements, distributed in an infinite number throughout its mass—the cerebral nerve-cells.

These lie in juxtaposition and enter into close relationship one with another. They are further arranged in regularly stratified zones one above another; and they form by their prolongations a tissue which is everywhere continuous, and thus produces unity of action between this multitude of isolated elements.

As physiological deductions, the following consequences spring from the considerations previously stated.

The cortical substance represents an immense instrument constituted of nervous elements, each gifted, it is true, with its proper individuality, and yet intimately connected one with another.

The series of cells arranged in stratified zones, and the connections of the different strata communicating one with another, imply the idea that the nervous activities of each zone may be isolatedly evoked; that they may be associated one with another; that they may be modified in passing from one region to another, according to the nature of the intermediary cells brought into play; that, in a word, nervous actions, like vibratory undulations, must propagate themselves through one point of contact after another, following the direction of the organic substance that underlies them, either transversely or vertically, from the superficial to the deep regions, and vice versâ.

On the other hand, as regards the physiological significance of certain zones, and the relation of each to the phenomena of sensation and motion, we may, by the laws of analogy, suppose that the sub-meningeal regions, principally occupied by the small cells, may be specially connected with the phenomena of sensation, while the deeper regions occupied by groups of large cells may be considered as the most important regions that give rise to motor phenomena.

In fact, in applying to this question the data which are acquired from the study of the spinal cord, and which show us, for example, that where there are small cells (posterior horns) the phenomena connected with sensibility take place, and where, on the contrary, there are large cells (anterior horns), motor impulses are developed—it is rational, I assert, to see physiological where there are morphological analogies, and to consider the sub-meningeal regions of the small cells of the cortical substance as being the natural sphere of the diffusion of general and special sensation, and therefore the great common reservoir of all the united sensibilities of the organism. And on the other hand, we may consider the deep zones as being the centres for preparing and emitting motor stimuli.

This mode of considering the cerebral cortex, in its totality, as an instrument essentially sensori-motor, conceived on the same plan as that of the sensori-motor instruments of the spinal cord, will permit of the formulation of certain new propositions on the subject of the evolution and intra-cerebral transformation of the phenomena of sensibility into motor reaction.

CHAPTER III.

THE WHITE SUBSTANCE OF THE BRAIN.

THE white substance of the brain is composed of a series of tubules in exact juxtaposition, and serving as it were as isolated conductors for each group of cells with which they are in connection, like the electric wire that carries the imponderable matter secreted by the pile with which it is united.

These nervous tubules, of which the general direction is perceptibly rectilinear, are, like the nerve-cells so closely related to them, distributed in very considerable numbers, seeing that they constitute the mass of the white cerebral substance of the two hemispheres.

They are essentially composed of a fundamental fibre designated as the *axis cylinder*, which represents the true nervous element of the tubule; and it is this fibril that usually enters into direct connection with the essential structure of the nerve-cell.

This fundamental fibril is surrounded, as with a muff, by a sheath of connective tissue, of variable thickness according as it is observed in the central or peripheral regions of the system. Between this sheath and the *axis cylinder* there is placed a fatty, oleo-phosphoric, highly refracting substance, the myeline, which forms, as it were, a fluid isolating body between these two elements.

THE WHITE SUBSTANCE OF THE BRAIN. 27

The neuroglia, with its thousands of meshes infinitely divided, similarly forms around the nerve-tubules a closely-woven network, which sustains them and con-

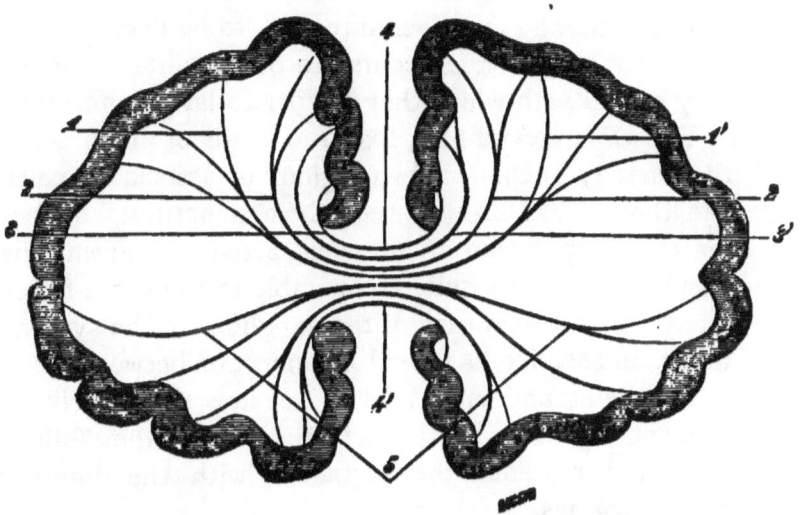

FIG. 3.—Diagram of the commissural fibres of the anterior regions of the brain. These form a series of curves one within another, the extremities of each of which plunge into the homologous regions of each cerebral lobe, 1, 1', 2, 2',—3 and 3'. They pass through the middle line, and at 4 and 4' give rise to the various appearances which the corpus callosum presents. 5. Commissural fibres of the inferior regions. These are curved in an inverse direction as regards the former, the convexity of each set being presented towards that of the other.

stitutes a uniting frame-work, and a veritable cement in the midst of which they are imbedded.

These thousands of nervous elements, thus constituted, emerge isolatedly from the different zones of the cortex, either directly, from the essential protoplasmic structure of the nerve-cells, or indirectly, by springing from the midst of the intercellular tissue, in the form of grey transparent fibrils, covered with an exceedingly delicate sheath. By degrees, in consequence of the interposition of the myeline, which becomes more abundant between the *cylinder* and the sheath, these grey fibrils assume the condition of white fibres, and, having attained the constitution of complete nervous elements, pursue their way in a given direction, to be decomposed, in the last stage of their course, in the satellite masses of grey matter with which they are particularly connected.

The white nerve fibres, like true bonds of union, serve then merely to connect two regions of associated cells, and thus to establish between them a natural channel for the propagation of nervous activity. From this standpoint they are quite comparable to the nerve-fibres interposed between each of the ganglions of the sympathetic, and serving as a bond of connection between them.

This being understood, let us see how these fibrillary elements behave, what particular direction they follow, and what relations they establish with the different central regions.

Generally speaking the white cerebral fibres take two directions.

1. The first group of the commissural fibres runs in a perceptibly tranverse direction.

Originating in the midst of the plexus of cells of the cortical substance, after having travelled with their partners for a while they separate from them one by one, abandon their primitive direction, pass across the mesial line, and are finally lost in the homologous regions of the opposite hemisphere. (Figs. 3 and 4.)

They thus constitute the transverse fibres of the vault of the corpus callosum; to which those of the anterior white commissure are attached.

They individually present themselves as curvi-

THE WHITE SUBSTANCE OF THE BRAIN. 29

linear fibres in the form of an U; and the branches of this U plunge in a similar manner into the homologous regions of both hemispheres.

This collection of transverse white fibres, which, taken as a whole, forms a little more than half the white mass of the cerebral hemispheres, establishes,

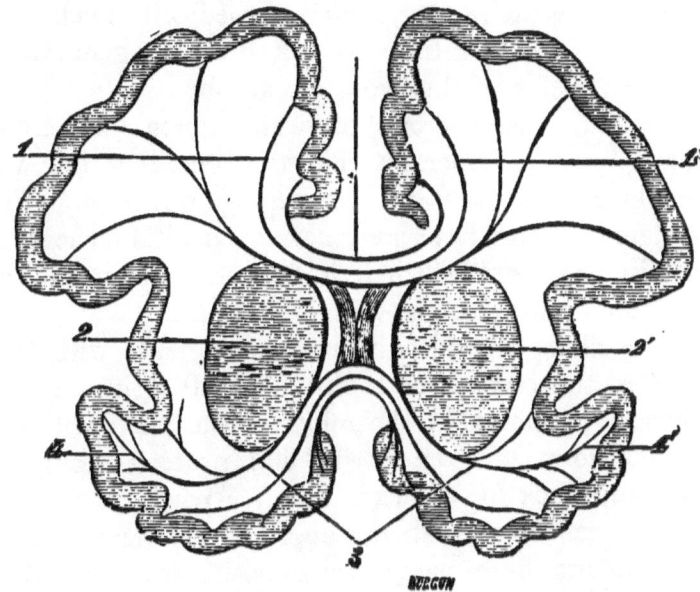

FIG. 4.—Diagram of the commissural fibres on the level of the corpus striatum—1, 1'. Groups of transverse fibres, one within another, continuous with those in the previous figure—2, 2'. Grey substance of corpus striatum—3'. Groups of inferior commissural fibres—4, 4'. These curve into the shape of an S to accommodate the corpus striatum, which they help to limit externally.

therefore, intimate connections between homologous regions of the cortical substance. The fibres themselves are thus, by reason of their relations with the grey elements, true commissures distributed everywhere in infinite numbers. We may also say that they constitute a very distinctly defined system of fibres, which by reason of its anatomical function may be in a general manner denominated a system of *commissural fibres*.

From a physiological point of view, on the direction of this order of fibres we might base the induction that it is by means of them that the regions of the two cerebral hemispheres are regularly anastomosed, cell to cell; and that they are, from this very fact, the true agents in the unity of action of the two cerebral lobes.

2. The second group of white fibres (converging fibres), no less important than the preceding, follows a rectilinear and sensibly converging direction. This system of fibres is entirely developed within the same hemisphere from which it is derived. It has nothing in common with the opposite hemisphere.

The fibres of which it consists originate with their fellows, the commissural fibres, at all points of the cortical periphery, in the midst of the plexus of cells, in the form of grey fibrils, and proceed along the common track for a certain time. Arrived at the level of the wall of the superior angle of the ventricles, the commissural fibres pass to the opposite side, while these insensibly approach one another like a series of rays radiating from the periphery of a hollow sphere, group themselves in the form of great white cylindroid fascicles placed in juxtaposition, and are inserted, like pins in a pincushion, around the anterior, middle, and posterior regions of the optic thalamus of the corresponding hemisphere.

By reason of the direction and special mode of grouping of the nervous elements which thus serve as a bond

THE WHITE SUBSTANCE OF THE BRAIN. 31

of union between the peripheral and central regions of the brain, we cannot but recognise that, anatomically, they play the part of converging elements and con-

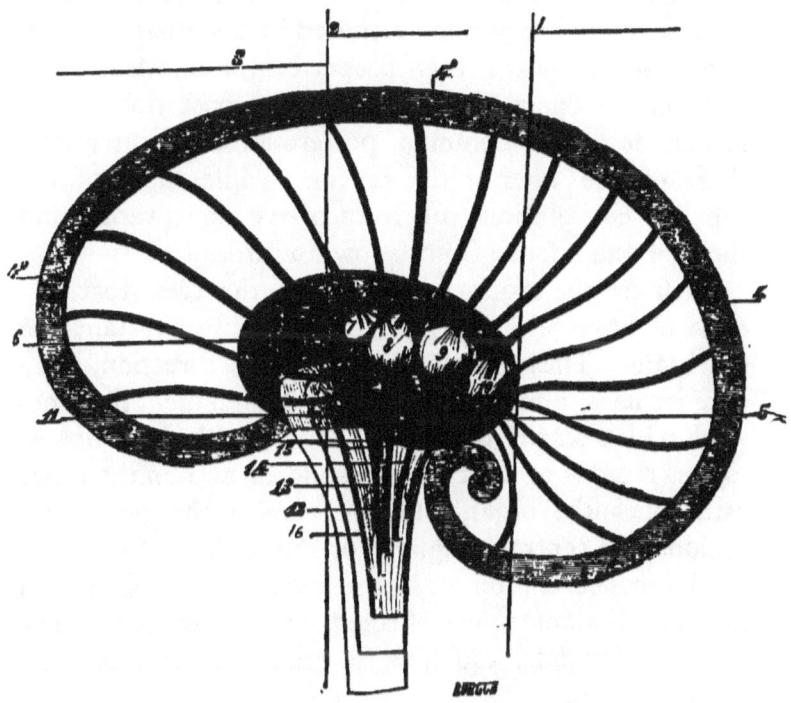

FIG. 5.—Diagram of the converging fibres and their relations to the central grey ganglions—1. Converging fibres of the posterior convolutions of the brain—2. Converging fibres of the middle convolutions of the brain—3. Converging fibres of the anterior convolutions of the brain—4, 4', 4". Cortical periphery as related to the central grey ganglions—5. Optic thalamus—6. Corpus striatum—7. Anterior (olfactory) centre—8. Middle (optic) centre—9. Median (sensitive) centre—10. Posterior (acoustic) centre—11. Central grey region—12. Ascending grey fibres of visceral innervation—13. Grey optic fibres—14. Ascending sensitive fibres—15. Ascending acoustic fibres—16. Series of anterolateral fibres of the axis going to be lost in the corpus striatum.

stitute a system, as well defined as the former, which we have described under the name of *converging fibres*.

As regards the behaviour of each group of converging

fibres, it is not our business in this work to give a detailed anatomic description of each of them. We shall only mention that, whether we consider them in the posterior, middle, or anterior regions of the brain, we find them everywhere disposed in a similar manner, and directed towards their proper centre of attraction.

Thus the converging fibres of the posterior convolutions follow a common postero-anterior direction; those of the anterior the reverse; while those of the superior convolution run from above downwards, and those of the inferior from below upwards.

Such are the special characters of the two great systems of fibres which constitute the white substance of the brain. These fibres run in a fixed direction, obey definite laws, and thus become the fundamental framework which on the one hand binds together the homologous regions of the two hemispheres, and on the other establishes the organic union between the peripheral regions and central ganglions of the brain.

This concentration of the converging fibres around the optic thalamus once effected, what becomes of these nervous elements, and how do they become lost in its mass?

From the moment in which they are implanted in the circumference of the optic thalamus, they become dispersed by degrees, insensibly taper away, and we then see them, in the form of whitish rectilinear fibrils, continue the converging direction of the primitive fascicles, and finally lose themselves in the midst of the different agglomerations of grey matter that they meet in their passage (centres of the optic thalamus and corpus striatum).

THE WHITE SUBSTANCE OF THE BRAIN.

Thus it is that each region of the cortical periphery is united, by means of these white fibres, to a symmetrical region in this common ganglion of grey matter (the optic thalamus), and that these two foci of nervous activity, the cortical periphery and the central ganglion, like two electric piles united by a common wire, are intimately united into a single instrument.

CHAPTER IV.

THE OPTIC THALAMUS.

HAVING thus passed in review the structure of the cortical substance, and the direction of the white fibres which emerge from it, it is now necessary to begin the study of the optic thalamus and corpus striatum, in the substance of which these white fibres are lost; these being, as it were, the natural pivots around which all the elements of the system gravitate.

The central mass of grey matter which is usually designated the optic thalamus, and of which the anatomical structure and general relations were scarcely known until the present day, is an ovoid body of reddish colour, situated in the very middle of the brain, a fact easily verifiable with a pair of compasses. It is in a manner the centre of attraction of all the fibres, the grouping and direction of which it thus governs.

It is composed: (1). Of a series of small isolated ganglions of grey matter, situated one behind another in a line which runs in an antero-posterior direction; (2). Of two slender bands of greyish material, lining the internal surface of the third ventricle, and continuous with the grey matter of the spinal cord, which thus ascends into the interior of the brain.

1. The isolated ganglions are four in number. These

THE OPTIC THALAMUS.

have already been described by anatomists, Arnold in particular,* with the exception of the median ganglion, the existence of which has been revealed by my own researches. They are arranged, as has been said, in an antero-posterior direction, and form successive tuberosities on the surface of the optic thalamus, which give it the multilobular appearance of a conglomerate ganglion. (See 7, 8, 9, 10, Figs. 5 and 6.)

The anterior ganglion is the most prominent. It is very much developed in the animal species in which the development of the olfactory nerve is very well-marked (*corpus album subrotundum* of anatomists).

Immediately behind comes a second, the middle ganglion, which in man is comparatively the most apparent and the most fully developed. In those animal species in which the optic nerves are rudimentary, the mole in particular, this ganglion is on the contrary scarcely visible.

Behind the preceding, and in the very centre of the optic thalamus, we meet with a third ganglion, of the size of a large pea, and whitish in appearance, which from its situation I propose to call the *median centre*.

Finally, behind, in the neighbourhood of the superior *tubercula quadrigemina*, we find another ganglion, of which the contours are in general vaguely defined, and which constitutes the posterior centre.†

By means of a series of sections, either vertical or horizontal, we may convince ourselves that these small ganglions form circumscribed and very distinctly iso-

* See "Tabulæ Anatomicæ."—Arnold, Icones cerebri et medullæ spinalis.' Turici, 1858.
† See "Iconographie photographique des centres nerveux." Plates, 2, 4, 6, 7, 8, 26, 28.

lated masses of grey matter, composed of plexuses of anastomosing cells; and that they in reality form small independent centres in regular juxtaposition, and isolatedly communicating with special groups of afferent nerve-fibres.

Now what is their true signification from a physiological point of view?

Up to the last few years the function of this mass of grey matter which forms the optic thalamus was an insoluble problem for anatomists. It was like an unknown land, of which anatomy had barely ascertained the situation. Thus, *à fortiori*, it may be comprehended that it was far from possible to point out the significance of each of these isolated ganglions.

It was by applying myself to the study of the connections of each of these little isolated centres with the peripheral nervous expansions which are distributed to them, and by confronting these new data with the facts which comparative and pathological anatomy had revealed to me, that I was led to consider them as so many small isolated and independent foci of concentration for the different kinds of sensorial impressions which are conveyed to their substance.*

Thus, if we take the anterior centre, for instance, (Fig. 6), we see that it is directly united, by means of a series of curvilinear fibrils, described by anatomists under the name of *tænia semicircularis*, with a particular mass of grey matter situated at the base of the brain, and itself directly receiving the external root of the olfactory nerve. Direct anatomical examination shows,

---

* See Luys, "Recherches anatomiques, physiologiques et pathologiques sur les centres nerveux," 1865.

THE OPTIC THALAMUS.

then, that there are intimate connections between the anterior centre and the peripheral olfactory apparatus. (20, Fig. 6.)

On the other hand, in confirmation of this, in the animal species in which the olfactory apparatus is very much developed, this ganglion itself is proportionally very well marked.

Analogy has thus led us to conclude that this ganglion is in direct connection with olfactory impressions, and that this marks it as the point of concentration towards which they converge before being radiated towards the cortical periphery.

This simple and purely anatomic view was, for myself, to some extent a ray of light, and the real clue that enabled me to propound my theory concerning the physiological function of the optic thalamus, by applying to the physiological interpretation of other ganglions the indisputable facts I had just ascertained respecting the anterior centre; since it was evident that what was true for one must be true for all the other closely related centres.

Thus by successively applying the same processes of investigation I arrived at the following conclusions:[*]—

That the middle centre, so manifestly in continuity, as regards its tissue, with the grey roots of the optic nerves, is destined for the condensation of visual impressions (Fig. 6—13. 14.); that the median ganglion is connected with the condensation of sensitive impressions (Fig. 6—8. 9.), and the posterior with that of

---

[*] I wish to remark, with regard to acoustic impressions, that the acoustic nerves, as they are implanted in the spinal axis, occupy precisely its most posterior regions, being situated behind the bundles of sensitive fibres.

auditory impressions (Fig. 6—3. 4.) ; and that thus, in their central order of classification, isolated sensorial impressions find independent halting-places grouped along the same line, and in an order correlatively similar to that which presides over their mode of distribution in the peripheral regions of the system.

It is, indeed, curious to observe in a human head, when examined in profile, that the olfactory organs of the nose are first met with, in the most anterior plane ; then the visual organs, the eyes, in the second line ; the sensitive organs in the third ; and finally the auditory organs, the ears, occupying the most posterior place ; and that, further, in their mode of distribution in the central ganglions of the cerebral mass these same impressions are grouped in isolated independent ganglions, occupying, as regards one another, a taxonomic order, which is in a manner only a repetition of their mode of origin in the peripheral regions (20, 13. 8. 3 ; Fig. 6).

These facts, which have shed quite a new light upon the anatomical and physiological functions of the optic thalami, have found their confirmation, on the one hand in the experiments of physiology, and on the other hand in the clinical examination of symptoms, which are in these matters the irrefragable criterion of every truly scientific doctrine.

Thus Dr. Edouard Fournié, in a series of experiments made on living animals by means of the injection of irritating substances into different parts of the optic thalamus, succeeded in annihilating such or such sensorial impressions, according as the traumatic laceration had attacked such or such a ganglion of the optic thalamus. Thus he succeeded in successively annihilating vision, sensation, smell, etc.*

On the other hand, well-observed clinical facts, reported by former writers, and therefore much anterior to my own researches, showed me that sometimes sensorial impressions might be totally and successively destroyed when the two optic thalami were simultaneously attacked, and that sometimes isolated sensorial impressions might be disturbed in consequence of a local lesion of their tissue.

There exists, indeed, a typical observation made by Hunter, of which he has left a drawing, which manifestly confirms what I have just put forward.

In this observation he recounts the curious history of a young woman who, in the space of three years, successively lost the senses of smell, sight, hearing, and sensation, and who gradually sank, remaining a stranger to all external impressions. When the autopsy of her brain was made, it was found that the optic thalami of each hemisphere, and the optic thalami alone (as can be seen in the original drawing)† were attacked by a fungus hæmatodes, which had progressively destroyed their substance.

In other circumstances, when circumscribed lesions have attacked separate ganglions, the abolition of such or such kinds of sensorial impressions has been noted. Thus, in three observations, for which I am indebted to the kindness of Dr. Auguste Voisin, and in which the abolition of smell on one side had been remarked, cor-

---

* " Recherches experimentals sur le fonctionnement du cerveau," par E. D. Fournié, Paris, 1873, p. 83.
† " Medico-Chir. Transactions"—London, 1825, vol. xiii. Part Second, p. 88.

responding degenerations of the anterior centres were likewise observed.

In a case reported by Serres, in a man who had suddenly lost the sight of both eyes, they found on autopsy a hæmorrhagic effusion occupying the optic thalamus on a level with the grey commissure, that is to say on a level with the middle centres.

In two unpublished cases observed by myself, I noticed a loss of sensation on one side of the body, coincident with an isolated destruction of the median centre of the opposite side.

Finally I have twice observed, in the brains of two deaf-mutes, in one case a lesion of the posterior regions, in the other amyloid degeneration of the same locality (posterior centres).* On the theory thus supported by the data furnished by normal and pathological anatomy, and experimental physiology, we may, therefore, legitimately conclude that the isolated ganglions of the optic thalamus are so many independent departments for each kind of sensorial impressions, and that the destruction of each of them may lead to the disappearance or alteration of the function to which it is specially devoted.†

2. The central region of grey matter which, as we have seen, lines the internal walls of the optic thalami, represents an elongation into the brain of the central grey matter of the spinal cord.

---

* See "Annales des maladies de l'oreille et du larynx," 1875. "Contributions à l'étude des lésions intra-cérébrales de la surdi-mutité," Luys.

† See the facts described in my 'Recherches sur l'anatomie, la physiologie, et la pathologie du système nerveux," p. 535, &c., as complementary details on the subject of the symptoms determined by different lesions of the optic thalami.

THE OPTIC THALAMUS. 41

It presents the appearance of two tracts of ashen-grey matter, which here and there form protuberances, which are themselves individually connected with the nerve-fibrils which are implanted in them. Such are the grey protuberances of the *septum*, for the internal olfactory roots; those of the *tuber cinereum*, for the optic fibres; the mamillary tubercles and pineal gland, for the connecting fibres emanating from the anterior centres.

It similarly receives a certain contingent of grey ascending fibres, which probably represent the centripetal spinal fibres which are distributed in these plexuses.*

The central grey matter is composed of a network of anastomosing cells, forming a continuous plexus.

Since, on the other hand, we can demonstrate that the white cerebral fibres radiating from the convolutions do not all lose themselves in the small centres of the optic thalamus, but that a certain number of them, pursuing their primitive direction, are prolonged as far as the plexuses of the central grey matter, we may legitimately recognize in this anatomical arrangement the natural channel for the propagation of nervous actions emanating from the cortical periphery, and manifesting themselves in the plexuses of central grey matter; and reciprocally, interpreting things in an inverse sense, we may recognize in this species of nerve-fibres the direct means of communication between the spheres where the phenomena of vegetative life take place, and those regions of the cortical substance which are the theatre of psycho-intellectual activity.

To sum up, the optic thalami are in a special manner

* See Luys' " Iconographie des centres nerveux." Plate 65.

the natural anatomical foci which preside over the organization and grouping of the cerebral fibres. From a physiological standpoint, the optic thalami are intermediary regions interposed between the purely reflex phenomena of the spinal cord and the activities of psychical life.

By their isolated and independent ganglions they serve as points of condensation for each order of sensorial impressions that finds in their network of cells a place of passage and a field for transformation. It is there that these are for the first time condensed, stored up and elaborated by the individual metabolic action of the elements that they disturb in their passage. It is thence, as from a penultimate stage, that, after having passed through ganglion after ganglion, along the centripetal conductors which transport them, they are launched forth into the different regions of the cortical periphery in a new form—*intellectualized* in some way, to serve as exciting materials for the activity of the cells of the cortical substance. (Fig. 6—14. 9. 4.)

These are, then, the sole and unique open gates by which all stimuli from without, destined to serve as *pabulum vitæ* for these same cortical cells, pass ; and the only means of communication by which the regions of psychical activity come into contact with the external world.*

* From the intimate connexions which unite the plexuses of the optic thalamus with those of the cortical layer, and which cause these latter, as regards the evoking of their activity, to be completely dependent upon the materials transmitted to them, it may be understood what an important part the morbid activity of the plexuses of the optic thalami may play in the evolution of various hallucinatory processes. See, for complementary details on the importance of irritations of the optic thalami in the development of hallucinations, the inaugural thesis of Dr. Ritti. Paris, 1874.

On the other hand, a direct examination of the relations of the centres of the optic thalami to the different regions of the cortical periphery enables us to determine the following peculiarities also.

It is sufficient to cast a glance over horizontal sections of the brain to recognize that each of these centres is more particularly in connection with certain regions of this very cortical substance. Thus, for instance, we see plainly that the central ganglion, by means of the white fibres that emerge from it, apparently radiates the impressions it condenses towards the antero-lateral regions of the brain,* and that the posterior centre acts in the same manner as regards the regions of the posterior *cornua;* while the median centre, by means of the divergent fibrils which are implanted in its mass, appears to direct its radiations indifferently towards all parts of the cortical substance. The anterior centre, less distinctly attached to the cortical substance, seems, nevertheless, to have its special area of distribution in the grey matter of the hippocampus. In the animal species in which the olfactory organs are well developed, this convolution similarly exhibits a high degree of development.

These anatomical data, which every one can observe, *de visu*, throw a completely new light upon that long-discussed question as to cerebral localizations, and are direct evidence that there are in the different regions of the cortical substance isolated circumscribed localities, affected in an independent manner, for the reception of such or such kinds of sensorial impressions. We are

* See my " Inconographie photographique des centres nerveux." Plates iv. v. vi.

thus logically led to comprehend that the peripheral development of such or such a sensory organ is designed to have a receptive organ in some way adapted to it in the central regions, and that the richness in nerve-elements of such or such a region of the cortical substance itself, and the degree of proper sensibility and specific energy of each of them, may, at a given moment, play an important part in the sum total of mental faculties, and thus determine the temperament of the specific activity of such or such an organization.

We thus recognize the fact that the secret of certain aptitudes—of such or such a native predisposition, is naturally derived from the preponderance of such or such a group of sensorial impressions, which find in the regions of psychical activity in which they are particularly elaborated a soil ready prepared, which amplifies and perfects them according to the richness and degree of vitality of the elements placed at their disposal.

Finally, the plexuses of the *central grey matter*, which are similarly united to the different regions of the cortical substance, show us that stimuli radiated from the depth of visceral life ascend, with the organic tissue which carries them, as far as the interior of the brain* (the experiments of Schiff confirm this); and that they are thus carried into the different regions of the cortical substance, and associated with the essential phenomena of psychical activity.

From this double induction we are therefore led to

* The experiments of Schiff tend to show that the vascular nerves of the liver and stomach pass over the *medulla oblongata* to terminate higher up. "A portion of them," he says, "appears to reach the optic thalamus." ("Compte rendu de l'académie des sciences," 15th Sept., 1862.)

THE OPTIC THALAMUS.

consider the masses of grey matter, usually described under the name of optic thalami, as essentially central regions which are the bond of union between the various elements of the entire cerebral system.

Through their tissues pass vibrations of all kinds, those which radiate from the external world, as well as those which emanate from vegetative life. There, in the midst of their cells, in the secret chambers of their peculiar activity, these vibrations are diffused, and make a preparatory halt; and thence they are darted out in all directions, in a new and already more *animalized* and more assimilable form, to afford food for the activity of the tissues of the cortical substance, which only live and work under the impulse of their stimulating excitement.

CHAPTER V.

THE CORPUS STRIATUM.*

THE mass of grey matter designated by the name of corpus striatum is the complement of the optic thalamus, with which it constitutes those two grey ganglions which occupy the central region of each hemisphere, and which are, as has been frequently pointed out, the natural poles around which all the nervous elements gravitate.

While the optic thalami present, in a manner, masses of grey matter grouped around the prolongation of the posterior columns of the spinal axis, of which, speaking in general terms, they form the crown, the corpora striata are, on the contrary, situated on the prolongation of the antero-lateral columns. They therefore evidently occupy an anterior situation as regards the optic thalami; and in connection with this subject, it is not without interest to remark that the same relations that exist in the whole of the spinal cord, are here reproduced with obviously analogous characteristics.

In the cord the sensitive or excito-motor regions occupy the posterior portion, while the essentially motor regions occupy the anterior.

In the brain the same relations as to neighbourhood,

* Fig. 6, p. 61, and Fig. 5, p. 31.

and the same correlative arrangements exist. Indeed, while the optic thalami with their different ganglions represent the regions of passage for sensorial impressions, the grey matter of the corpora striata, with its multiple elements, represents the place of halt and reinforcement for motor stimuli radiating from the cerebral cortex.

It may therefore be said that in the brain, by virtue of the same anatomic arrangements, the regions where the phenomena of sensation occur, and those in which motor stimuli are elaborated, reciprocally maintain the same topographic relations that they have in the different portions of the spinal cord proper.

As to external configuration the mass, of the corpus striatum presents the form of a reddish grey mass, of flabby consistence*, situated in front of the optic thalamus, and, gradually diminishing from before backwards, extending as far as its posterior regions.

It follows that the mass of the corpora striata presents an ovoid pyriform appearance, the larger extremity directed forwards, the tapered extremity backwards, and that the optic thalamus in its antero-lateral regions is encircled with a network of grey substance having its maximum thickness anteriorly.

It follows besides, as a consequence of this anatomical arrangement, that the white converging fibres which group themselves around the optic thalamus, before arriving at their destination encounter a more

* This soft consistence of the grey matter of the corpus striatum, on the one hand, and the excessive development of the capillaries which circulate in its mass on the other, explain the extreme friability of its tissue and the facility with which it may be lacerated by congestions which render the vessels turgid. Thus we may account for the greater frequency of unilateral paralyses of motion, as compared with those of sensation.

or less considerable thickness of the corpus striatum, which they traverse from one side to the other, at various angles and in various directions. (Fig. 5.) The anterior convergents in particular, which run towards the corresponding regions of each optic thalamus, plunge from before backwards into the very mass of the corpus striatum, and divide it into two segments, one extra- and one intra-ventricular.*

The colour of the grey matter of the corpus striatum is sensibly homogeneous, wherever it is observed. It is flabby, reddish, and composed of special anatomical elements. It is, moreover, permeated by an infinite number of whitish serpentine filaments, which represent the terminal expansions of the antero-lateral motor fibres of the spinal cord.

In the internal and inferior regions, however, where there is a confluence of all the antero-lateral fascicles of the spinal axis which expand into the corpus striatum, we come upon a very clearly circumscribed region of firmer consistence, and yellowish colour, which is easily recognized by its peculiar striation.†

This peculiar circumscribed mass of yellowish matter, which I have more particularly designated under the name of *yellow nucleus* of the corpus striatum, plays an important part, as a centre of radiation for nerve-fibres, in its relations with the ultimate expansions of the cerebellar peduncles.

* These are those fibres composed of isolated fascicles regularly stratified one above the other, of which the continuity may be clearly ascertained by the aid of dissection, in antero-posterior sections of the brain. They are most improperly designated by the name of the *internal capsule*. (*See* "Iconographie photographique," pl. xlv.)

† See "Iconographie photographique," pl. xxxi. xxxii. x. xi. xliii. and xliv.

THE CORPUS STRIATUM.

The structure of the corpus striatum must now be considered, as regards:

1. The study of the grey matter, regarded from a histological point of view, and as to the characters of its elements.

2. That of the nervous elements, of various origin, which enter into relation with those proper to itself.

1. The grey matter of the corpus striatum is histologically composed of an infinite number of large polygonal nerve-cells with multiple prolongations, their size being in general about the same as that of the larger cells of the cerebral cortex. These cells, considered in themselves, present characters common to all the other cells. They are provided with what appear to be a nucleus and nucleolus, and present ramified prolongations which rapidly taper away, and constitute with those of the neighbouring cells a very dense and very delicate network.

Besides these large cells just mentioned, we also meet with elements of smaller size, especially in the region of the yellow nuclei, where they are extremely abundant. Their histological characters recall in a more or less vivid manner the similar elements met with in the deep zones of the grey matter of the cerebellar convolutions. These small elements, of which the nucleus is voluminous, and of which the yellowish colour enables us to distinguish them from the surrounding corpuscles of the neuroglia, exhibit a fringe of radicles of extreme tenuity, which is lost in the network formed by the large cells. It seems, then, probable that these small cells, which to some extent histologically represent the cerebellar element, enter more or less directly into combination with the radiations from the large cells which represent the cerebral element.

Besides these two principal elements, we have still to describe the corpuscles of neuroglia, derived more or less directly from the sheaths of the capillaries, and a considerable number of vessels which directly penetrate from below upwards into its interior in the form of more or less rectilinear filaments. (Perforated space of Vicq d'Azyr.)

2. The diverse elements which enter into the anatomic constitution of the corpus striatum, are divided into two special groups: 1, some may be considered as a system of fibres afferent to the corpus striatum; 2, others as a system of efferent fibres.

1. The first group comprehends: *a*, on the one hand, all the cerebral fibres radiating from the different regions of the *cortex*, and lost in the substance of the corpus striatum (cortico-striate fibres); β, on the other hand the ultimate expansions of the superior cerebral peduncles, which are lost in its mass, and which represent the specific importation of the cerebellar element into the constitution of motor phenomena.

a. (Fig. 6—6. 11. 16.) The elements of the first group, which, on account of their origin and termination, I have proposed to call cortico-striate, belong to that mass of convergent fibres which, radiating from all points of the cortical periphery, and probably from the psycho-motor regions so clearly determined at the present day, take a common direction towards the central ganglions. This order of fibres, however, once arrived at the circumference of the optic thalamus, instead of terminating like their fellows, only embrace it. Arrived at the frontier of

THE CORPUS STRIATUM. 51

the optic thalami and the corpora striata, these fibres are immediately reflected from below upwards, in the form of spiroid lines, and are finally isolatedly distributed in the different cell-territories of the corpus striatum with which they are especially connected.

These cortico-striate fibres, which have come out of the depths of the cortical layer with the sensitive fibres, still proceed for a certain distance through the brain, in juxtaposition with these latter, as is also the case in the peripheral nervous trunks, which are composed of both sensitive and motor fibres, embraced in the same envelope. Soon, having arrived in the presence of their respective centres of attraction, they each obey their innate affinities, and are distributed, some to the centres of the optic thalamus, others to the different regions of the grey substance of the corpus striatum.

These fibres then represent, properly speaking, the natural bonds of union between the regions of the cortical substance whence proceed the different voluntary stimuli (psycho-motor centres), and the different cell-territories of the corpus striatum where they are reinforced. As regards anatomical analogies, they represent the whole group of anterior roots in its relations with the grey elements of the spinal cord.

Their precise origin in the different regions of the cortical periphery is still a problem to be solved for each of them in particular. (Fig. 6—6. 11. 16.) This is also the case as regards their central distribution in the different cell-territories of the corpus striatum. (2. 12. 17—Fig. 6.) At the present day they are only known and anatomically demonstrable in an intermediate portion of their transit, at the moment when they are reflected in the form of serpentine fibres;* and yet their existence, as centrifugal conductors of motor stimuli, radiating from the excitable zones of the cerebral cortex, is very clearly demonstrated. This is one of the most interesting points that experimental physiology has brought to light in recent times.

The recent researches of Fritsch and Hitzig, who were the first, in 1870, to point out that certain zones of the grey cortex were excitable by galvanic currents,† have opened the road in this direction.‡ Ferrier has shown, indeed, after them, that by applying electric excitement in such or such a region of the grey cortex, motor reactions in such or such isolated groups of muscles are determined; that at will we may cause the eyes, tongue, neck, etc., of an animal to move, according as we electrify such or such a convolution; and that, in a word, there are in the tissue of the cortical layer a series of small independent motor centres, which may be isolatedly excited, and which communicate by independent conductors with the different portions of the muscular system. More than this, it has been proved that things take place in the same manner in man; for an American physician, pushing the boldness of experiment to its ultimate limits, obtained similar results in a patient whose brain was denuded by a degeneration of the cranial case.§

* See " Iconographie photographique," pl. xxxii. and xxxiv.
† "Archiv of du Bois-Reymond," (1870, Heft. 3).
‡ See "Progrès médical," number 28, 1873, and number 1, 1874. "Recherches expérimentales sur la physiologie et la pathologie cérébrales," by Ferrier.
§ See the "Movement médical," 1874, number 33, pl. 381. "Recherches expérimentales sur les fonctions du cerveau humain," by Robert Bartholow, Professor in the Medical College of Ohio.

THE CORPUS STRIATUM. 53

Finally, in certain pathological cases in which I made special researches, still unpublished, I have even been able to demonstrate, as a proof of the existence in the cortical layer of isolated foci of motor excitation, that in persons who had undergone amputations at a distant date—subjects who had been long deprived of an upper limb, in the case of disarticulation of the shoulder, for instance—there existed in certain long disused regions of the brain, coincident, very distinctly localized atrophies. I have, moreover, demonstrated that the atrophied regions of the brain are not the same in the case of the amputation of a leg, and in that of amputation of the upper limbs.

These facts, then, as I have already explained in former works, already extending over ten years, authorize us to conclude that there exists a special order of nerve-fibres radiating from different departments of the cortical substance, and distributed in isolated territories of the grey matter of the corpus striatum, which is thus associated as a co-operant factor in all the vibrations that take place in the plexuses of cerebral cells; and to consider it as proved that these different groups of cortico-striate fibres have each an independent point of origin.

β. The afferent elements of the second group, as we have already indicated, are represented by the terminal expansions of the cerebellar peduncles.

The superior cerebellar peduncles, in fact, after intercrossing in the median line, become associated and form two masses of grey matter, described by Stilling, and recognizable by their reddish colour.

These two ganglions, which, as regards their structure

and connections represent a veritable focus of radiation for cells and nerve-fibres, give rise throughout all their antero-external substance to a series of fibrils, interlaced in a thousand ways, which all terminate in the form of yellowish filaments, in the grey matter of the corpus striatum. It is this special contingent—an indirect emanation from the active elements of the cerebellum—that gives to this particular department of the corpus striatum, that characteristic yellowish colour, which I have already described under the name of the yellow nucleus of the corpus striatum.

These fibrils of cerebellar origin which are disposed in the form of yellowish rayed filaments, taper away insensibly, and embrace the white spinal fibres which expand in the corresponding regions of the corpus striatum; and are probably lost in the network of large cells, as has been previously suggested.

Now, how do they terminate? What is the ultimate mode of combination of the individual elements which represent in the brain the activity of the cerebellum? How does the cell of the corpus striatum come into contact with the cerebellar elements of the new importation?

So far I have only been able to form conjectures, and while supposing that there must be some sort of anatomic combination between these elements of varied origin, I can only pause and await the results of future researches.

However it may be, we cannot help considering the corpus striatum, from a dynamic point of view, as being indirectly connected with the phenomena of cerebellar activity, and seeing in the superior cerebellar peduncles,

THE CORPUS STRIATUM.   55

in the red ganglions of Stilling, and in the yellowish radii which emerge from them, so many centrifugal conductors, incessantly active foci of nervous radiation, which allow the cerebellar motor influences with which they are charged to overflow into these plexuses.

The cerebellar innervation, is thus intimately associated with the vital phenomena of the corpus striatum as a true auxiliary force. It is incessantly overflowing into its thousand plexuses like a continuous current of electric force, and, as it were, charges its nerve-cells. In motor phenomena it is associated with all our different motor acts, and gives to our movements their regularity, their force, and their continuity. Under a thousand forms, in fact, it silently disperses itself through all the conscious and unconscious actions of the organism, and seems to be an indispensable component of every motor act whatsoever.*.

2. The elements of the second group, those which constitute the mass of the *efferent* fibres of the corpus striatum, are represented by that series of nerve-fibres which are ordinarily described under the name of cerebral peduncles, and which, grouped in the form of isolated fascicles, and arranged in a spiroid fashion, pass in succession, after having traversed the *pons*, to be dispersed in the different segments of the spinal axis. These fibres, which represent conductors interposed

---

* The researches of experimental physiology, as well as clinical phenomena, demonstrate in an unmistakable manner, the important part played by cerebellar innervation in motor acts. When the cerebellum is engaged in any way, disturbances in the regular co-ordination of movements are always perceived, and more than this, motor acts are weakened, a phenomenon of the utmost importance, as implying the extinction of the foci of motor innervation which are designed to produce them.

between the different cell-territories of the corpus striatum and the different ganglions of the motor nerves of the spinal chord, are not distinctly isolated at their point of origin in the plexuses of cells of the corpus striatum. (Fig. 6.—12. 12′. 17 and 19.)

All that we can say of them is, that they appear by insensible degrees in the form of whitish traces creeping over the grey matter of the extra- and intra-ventricular ganglions; that soon they insensibly increase in volume; that converging like the rays of a fan, they all approach the yellow nuclei of the corpus striatum; that they gradually enter into contact with the yellow fibres which constitute the substance of these bodies; and that when, after condensation, they emerge from the corpus striatum, they present themselves in the form of three demi-cones, one enclosing the other. (Fig. 6.—19.)*

These nervous elements, having been thus arranged and reinforced by the union of different masses of grey matter belonging to the cerebellar innervation (grey matter of Sömmering, grey matter of the *pons*) (Fig. 6.—18. 19. 19′. 19″) pursue a descending and oblique course, which causes them (on a level with the *medulla*) to pass insensibly into the opposite regions of the spinal axis. Little by little, and fascicle by fascicle, they separate, to distribute themselves in the different segments of the spinal cord, and in the different groups of motor cells of the antero-lateral regions. These, regularly stratified one above another, like a series of electric machines always ready to start into action,

---

* In a horizontal section of this region, these three demi-cones present the appearance of three semicircular concentric lines. *See* " Iconographie photographique," pl. xi. and xxxi., fig. 1.

silently await the arrival of the stimulating spark destined to call them into activity.

Thus it follows, from what we have just explained, that the corpus striatum, like the optic thalamus, is a nervous apparatus with multiform activities.

It is a common territory into which the cerebral, cerebellar, and spinal activities come in succession, to be combined, and I might almost say, to anastomose. It thus represents, from a dynamic point of view, a synthesis of multiple elements.

It is in the midst of its tissues that the influence of volition is first received at the moment when it emerges from the depths of the psycho-motor centres of the cerebral cortex. There it makes its first halt in its descending evolution, and enters into a more intimate relation with the organic substratum destined to produce its external manifestations—in one word, *materializes* itself. (Fig. 6.—12 and 17.)

From this moment it comes into intimate contact with the innervation radiating from the cerebellum, and it is now no longer itself, no longer the simple purely psycho-motor stimulus it was at its origin. It is associated with this new influence, which gives it somatic force and continuity of action. It then passes out of the brain by means of the peduncular fibres, combined with a new element, and pursuing its centrifugal course, it is finally extinguished here and there by setting in motion the different groups of cells of the spinal axis, whose dynamic properties it thus evokes. (Fig. 6.—18 and 19.)

Thus also, proceeding like an electric current into the different departments it animates, it now tends to

58   THE BRAIN AND ITS FUNCTIONS.

produce phono-motor movements designed to express outwardly the emotions of our sentient personality, and now to determine in the different muscular groups, general or partial movements of flexion, extension, or progression, according as it is distributed to such or such groups of satellite cells, the habitual servants of its excito-motor demands.

We see then, to sum up, by means of this simple physiological sketch, what an all-important part these two central ganglions play in the phenomena of cerebral activity, and how completely different is the mode of action of each.

The elements of the optic thalami purify and transform by their peculiar metabolic action impressions radiating from without, which they launch in an intellectualized form towards the different regions of the cortical substance. The elements of the corpus striatum, on the contrary, have an inverse influence upon the stimuli starting from these same regions of the cortical substance. They absorb, condense, *materialize* them by their intervention; and, having amplified and incorporated them more and more with the organism, they project them in a new form in the direction of the different motor ganglions of the spinal axis, where they thus become one of the multiple stimulations destined to bring the muscular fibre into play.

CHAPTER VI.

PHYSIOLOGICAL DEDUCTIONS.

Now, if we group synthetically the anatomical propositions we have tried to establish in the course of this work, we see that the brain is a geminate organ, formed of two hemispheres, of which the elements are strictly associated with one another, by means of a series of commissural fibres which unite them intimately, and produce a certain tendency in their molecules to vibrate in unison. (Figs. 3. and 4.)

Each of these two lobes, or hemispheres, is fundamentally formed of masses of grey matter irregularly distributed—the grey matter of the central ganglions (the optic thalami and corpora striata) and that of the cerebral cortex.

These two regions of cerebral activity are united to one another by a series of white fibres, which serve as a bond of union between them, and as a channel of propagation for nervous currents passing from one to the other, either centrifugally or centripetally.

The opto-striate central ganglions of each lobe may be ideally conceived as occupying the centre of a hollow sphere, of which the circumference is represented by the undulations of the cerebral cortex; and the white fibres would thus represent an infinite number of radii uniting the central with the peripheral regions of the sphere. (Fig. 5.)

The anatomical study which we have just made, of the grey matter of the optic thalamus and that of the corpus striatum, has enabled us to observe distinct differences between them, and consequently to formulate the unlike dynamic aptitudes with which each of these two ganglions is gifted.

We have thus seen that the function of the optic thalamus in particular seems to be that of receiving, condensing, and transforming, like a true nervous ganglion, impressions radiating from the sensorial periphery, before launching them into the different regions of the cortical substance; and that, inversely (Fig. 6.—14, 9. 4), the corpus striatum, in connection with exclusively motor regions, appears to be a place of passage and reinforcement for stimuli radiating from the different psycho-motor zones of the cortical periphery.

These anatomical connections being admitted as fundamental *data*, as regards the structure and mode of agency of the nervous elements; let us now see what use we may make of this, from the standpoint of the particular interpretation of certain phenomena of cerebral activity.

Let us take things as they normally occur, following the natural channels by which excitations from the external world penetrate into the organism. Let us take, for example, the impression upon a sensitive nerve —a vibratory phenomenon which calls into activity the cells of the retina or those of the acoustic nerves; what then takes place in the secret recesses of the nervous conductors?

PHYSIOLOGICAL DEDUCTIONS.

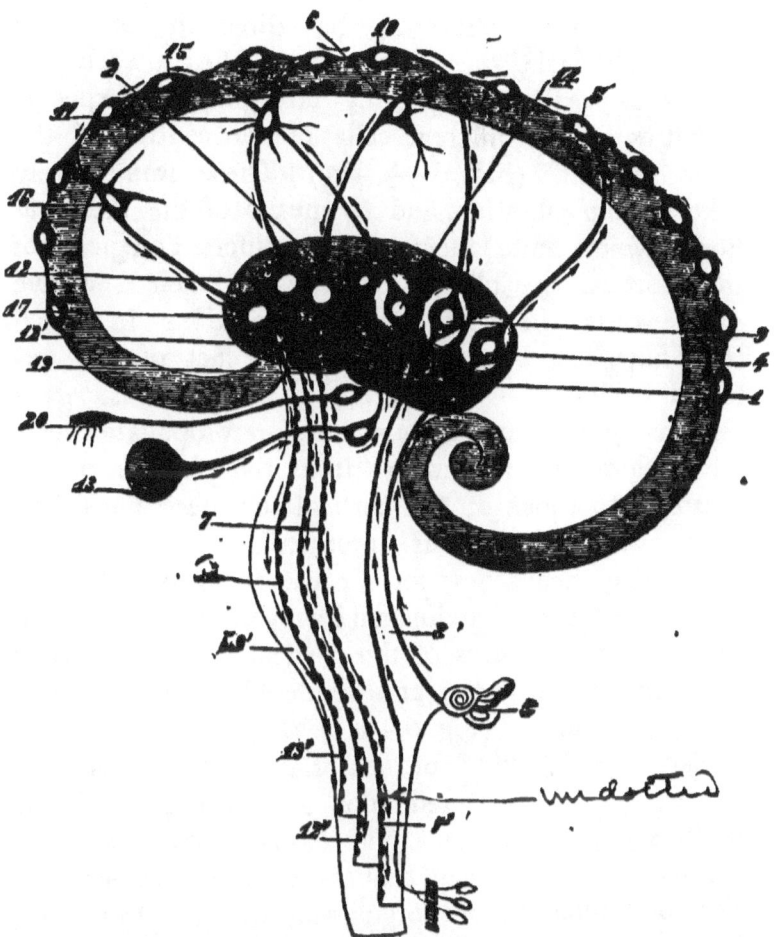

FIG. 6. Diagram of the sensori-motor processes of cerebral activity.—1. *Optic thalamus* with its centres and ganglionic cells—2. *Corpus striatum*—3. Course of the propagation of acoustic impressions. These arrive in the corresponding centre (4), are radiated towards the *sensorium* (5), and reflected at 6 and 6', to the large cells of the *corpus striatum*, and thence at 7 and 7', towards the motor regions of the spinal axis—8. Course of sensitive impressions. These are concentrated (at 9) in the corresponding centre—radiated thence into the plexuses of the *sensorium* (10), reflected to the large cortical cells (11), and thence propagated to the large cells of the *corpus striatum*, and finally to the different segments of the spinal axis.—13. Course of optic impressions. These are concentrated (at 14) in their corresponding centre, then radiated towards the *sensorium* (at 15). They are reflected towards the large cells of the *corpus striatum* and afterwards propagated to the different segments of the spinal axis; 18, 19', 19'', the antero-lateral fibres from their point of origin in the *corpus striatum*, are invested by the elements of cerebellar innervation which begin to appear in the peduncles (19), to become considerably thicker at 19', on a level with the region called the *pons* and to diminish insensibly on a level with the medullary regions, 19'.—20. Peripheral expansion of the olfactory nerves.

THE BRAIN AND ITS FUNCTIONS.

Immediately' following the direction of their natural channels, these vibrations applied to each particular sensorial nerve, bring into play the specific activities of the different cells of the centres of the optic thalami. (Fig. 6.—3, 13.) These immediately take up the vibration, and by means of the radiating fibres which unite them to the different regions of the cortical periphery, transmit to their sensitive partner-cells, the new dynamic conditions in which they have just been placed by the fact of the external excitation.—(See Fig. 6.—5 and 15.) External sensorial impressions do not therefore propagate themselves through and through from the plexuses of the sensorial to those of the cortical periphery, until they have awakened various intermediate cell-territories which give them a new form, cause them to undergo a peculiar metabolic action, and only launch them into the different plexuses of the cortical zones, after they have animalized them, and rendered them somehow more assimilable. (Fig. 6.—4. 9. 14.)

Each special kind of sensorial excitation is thus dispersed, and quartered upon a special area of the periphery of the brain. (Fig. 6—15 and 5.)

Anatomy shows, then, that there are definite localizations of limited regions, organically designed to receive, to condense, and to transform such or such particular kinds of sensorial impressions.

Experimental physiology has proved on its side, that in living animals, as the beautiful experiments of Flourens long ago showed, it is possible, by methodically removing successive slices of the cerebral substance, to cause these animals reciprocally to lose either the faculty of perceiving visual, or that of perceiving auditory impressions.*

More than this, Schiff, in his recent experiments, as ingeniously contrived as delicately executed, succeeded in demonstrating in a precise manner, that in the animal under experiment, the cerebral substance was subject to local increase in temperature, according as it was successively excited by such or such kinds of sensorial impressions; and that thus, in the brain of a dog, which was made to hear unexpected sounds, such or such a region of the cortical substance was heated, and that in another, in which tactile, olfactory, or gustative sensation was excited, other regions of the brain were reciprocally erethised and heated in an isolated manner.†

Following up the process of the migration of sensorial excitement from the peripheral to the central regions of the system, we see that all sensorial impressions arrive, in the last stage of their transit, at the plexuses of the cortical substance; that they arrive transformed by the action of the intermediate media through which they have passed *in transitu*; and finally that they there die away and are extinguished, to revive under a new form, by bringing into play the regions of *psychic* activity where they are at last received.

As soon as the sensorial excitation is dispersed in the midst of the plexuses of the cerebral cortex, new phenomena unfold themselves.

Here mere analogy leads us to think that the sen-

* Flourens "Recherches expérimentales sur le système nerveux." Second edition, 1842.
† Schiff, "Archives de physiologie," 1870.

sitive cells of the brain may behave like those of the spinal cord, and that in the presence of the physiological excitations proper to them, they will react in a similar fashion. We may, therefore, suppose that at the moment when the cerebral cell receives the impregnation of the external impression, it becomes *erect*, as it were, develops its peculiar sensibility, and disengages the specific energies which it contains.

Thus it is that the impression which is communicated and which manifests itself by a development of heat in certain regions of the cortex (as in the experiments of Schiff), is propagated through the circumjacent plexuses, and, according to the laws of undulatory movement, develops by degrees the latent activities of new groups of satellite cells, which in their turn become new foci of activity for the neighbouring cells, with which they are intimately anastomosed.

In this manner we can conceive how, in consequence of a simple sensorial impression, all the agglomerations of nervous elements of which the cerebral cortex is composed, may isolatedly become successively engaged; how the movement is communicated from point to point (Fig. 6—5. 10. 15.); how the individual sensibility of the nervous elements begins to take part in the phenomenon; how life is awakened in regions at first silent; and how, in a similar manner, the incident excitation, after having thrown into agitation different zones of the cortical substance, is finally transformed into a centrifugal excitation, *reflected*, and externally discharged in the form of a motor act. (Fig. 6—6. 11. 16.)

Having followed step by step, the phenomena of

cerebral activity just explained, and interpreted them in ordinary language, we may conclude that sensorial excitations radiated from the periphery reach the regions of psychic activity, and that there, coming under the influence of the elements of which it is composed, they become transformed into persistent impressions—ideas corresponding to their origin; that they bring into play the sensibility and emotivity proper to these regions; that they become associated, anastomose one with another in a thousand ways, by means of the organic tissue through which they are evolved; that they are amplified and transformed by the different zones of cells through which they are sifted; and that finally, they are exported and reflected outwards in the form of voluntary motor manifestations, expressions more or less indirect of a primordial phenomenon of sensibility.

Now, from the premisses of the structure of the cortical substance, comprehended as already indicated, it may be possible to deduce *data* which will enable us to appreciate the dynamic functions of the different zones of cells contained in it.

We have already established, that the elements which compose it have very distinct morphological characters; that the zones of small cells occupy the sub-meningeal regions, and that the zones of large cells occupy the deep regions. In the minute constitution of the spinal cord we find similar appearances as regards the distribution of the nervous elements; and we further know that the regions of small cells are the seat of sensitive, those of large cells the point of departure of motor, phenomena. The laws of analogy therefore lead us to suppose that morphological imply physio-

F

logical analogies, and that in the succession of the multiple activities of the cortical substance, we may probably suppose that the sub-meningeal regions, occupied by small cells, are more particularly the regions fitted for the reception of sensitive impressions, while the deeper layers, occupied by the large cells, appear to be more particularly centres of emission appropriated to motor phenomena.

This granted, we arrived at the following conclusion: That in the plexuses of the cortical substance, there is in those formed by the small cells a special sphere for the dissemination and reception of sensitive impressions, which all impinge here and bring into play the peculiar sensibility of the cells ; and that these zones, which are anatomically demonstrable, and which represent the posterior sensitive regions of the spinal cord, receive in their essential structure all the particular sensibilities of the organism, and produce a union between them.

They thus form that *matrix*, that *regio princeps* of the cerebral cortex, which constitutes the true *sensorium commune*, the common reservoir into which all the impressions that have thrown our sensitive fibres into agitation, flow, and in which they subside. (Fig. 1—B, and Fig. 6—5, 10, 15.)

Thus, then, is constituted this region which receives into its sensitive tissue the resultant of all sensitive excitations, from the external world as well as from the vegetative life of the organism, and which, when thrown into agitation, sensitized in its turn, reacts in a thousand ways, dispersing in all directions the vibratory excitations which have developed the energies of its elements. It is its task gradually to transform the phenomena of sensibility, and finally to cause sensitive excitations, radiated from cell-plexus to cell-plexus, like a force in evolution which is incessantly transformed, to produce insensibly a motor phenomenon.

Thus we arrive at the conclusion that, from a physiological point of view, the voluntary motor act which emanates from the brain is in all cases nothing more than the repercussion, more or less distant, of a primordial sensitive impression. (Fig. 6—6, 11, 16.)

It should, however, be added, that although the act of voluntary motion is, as a general rule, only the indirect expression of the agitation of the *sensorium*, nevertheless, from the very fact that it is evolved throughout the plexuses of the cortical substance, laying its various zones under contribution, it is not a simple, purely reflex phenomenon, like those which occur in the similar plexuses of the spinal axis; it is a complex synthetic phenomenon that resumes in itself the different elements which, taken together, constitute human personality. We may also say that if an act of the will be merely a phenomenon of transformed sensibility, it is, nevertheless, sensibility doubled, multiplied by all the cerebral activities in agitation, in a word, by the feeling and vibrating human personality, which comes into play in a somatic form, and reveals itself externally by a series of *reflected* and co-ordinated manifestations.

CHAPTER VII.

PHYSICO-CHEMICAL PHENOMENA OF CEREBRAL ACTIVITY.

THE nerve-cells, considered as to their intrinsic properties, individually participate in all the general phenomena of the life of cells. Like all their fellows they have their history, their genealogy, their periods of growth and decay. They are subject to alternate phases of repose and labour, and, like them, are all gifted with a specific histological sensibility which gives them special dynamic characters.

It is the blood alone that makes them live and feel; it is it alone which, as sole agent of their incessant activity, percolates everywhere through the nervous tissue, and carries with it the elements of all life and all movement.

This is so true, that if we succeed in momentarily suspending the circulation in the encephalon, the whole vital machinery stops at once, and every phenomenon of nervous activity is at the same instant interrupted.

Decapitated animals are by this very fact deprived of all cerebral functionment, and, it is a very remarkable fact, that if we succeed in artificially restoring to the cerebral tissue the nutritive materials of which it was deprived; if, by means of injections of defibrinated blood, such as Brown-Sequard has experimented with,

PHENOMENA OF CEREBRAL ACTIVITY. 69

we succeed in giving their habitual stimulation to the nerve-cells, the signs of life come back as if by enchantment, and the head of a dog, thus momentarily revived, will still afford ephemeral manifestations of a conscious perception of external things.*

In man the more or less complete arrest of the blood in the brain, produces accidents which are sometimes overwhelming, faintings, and loss of consciousness with stupor; and it is now recognized, thanks to the labours of modern physiology, that the greater number of those apoplectiform seizures, which were formerly attributed to a sanguineous plethora in the plexuses of the brain, should on the contrary be ascribed to a more or less complete arrest of the course of the blood in the capillary plexus. The attacks observed in these circumstances may be legitimately attributed to a sort of asphyxia of certain regions of nerve-cells (principally those of the *sensorium*, when we have to do with losses of consciousness, vertigos, and fainting-fits); the nervous elements being stupefied for an instant, in consequence of the more or less complete suspension of the arrival of their nutritive materials.†

The continuity of the sanguine irrigation is, then, the *sine qua non* of the regular working of the cerebral cells. It is at the expense of the juices exhaled from the walls

---

* Brown-Sequard once injected the head of a dog when separated from the trunk with defibrinated and oxygenated blood, and at the moment when the injection of this blood had recalled the manifestations of life, he called the dog by his name. The eyes of the head thus separated from the trunk turned towards him, as if the voice of the master had still been heard and recognized. (Annales "Médico-physiol.," 1870, p. 350.)

† Every one knows that in fainting fits, and syncope, the most rapid method of bringing them to an end is to place the patient horizontally, so as to favour the mechanical afflux of blood to the brain.

of the capillaries, that they feed themselves and continually repair the losses sustained by their integral constitution.*

Plunged into the midst of this humid atmosphere surcharged with phosphates, of which the materials are incessantly renewed, they extract from this vivifying medium the elements of their reconstitution ; like living beings plunged into the terrestrial atmosphere, borrowing from the surrounding air the *pabulum vitæ* which enables them to live, and sustains them. Thus it is that they successfully endure their expenditure of the phosphorated element during the period of their diurnal activity, and that they can maintain themselves in equilibrium as regards their receipts and expenditure.

This truth was very clearly brought to light by the ingenious researches of Byasson, who has pertinently shown that every cerebral cell in functioning expends its phosphorized materials, and that this waste resulting from cerebral activity, like the natural physiological excretions, is drained away from the organism by passing out in the urine, in the form of sulphates and phosphates, which serve as a chemical measure of the intensity of cerebral work done in a given time.†

* The blood which comes to the brain red and oxygenated, returns by the capillaries black and charged with carbonic acid. (Gavarret, "Phénomènes physiques de la vie." Annales médico-psychol. 1870, p. 347.)

† To arrive at these results, Byasson for several days submitted to a special physical and moral regimen. He estimated exactly the quantity of phosphates and sulphates which entered into his diet, and also the quantity excreted. At the end of a certain time, these fundamental data having been ascertained, he began to work his brain, and in proportion to the amount of his work, the diet remaining constant, the quantity of sulphates and phosphates excreted by the urine had increased in a notable manner. (Byasson, "Essai sur la relation qui existe à l'état physiologique entre l'activité cérébral et la composition des urines." Journ. d'anat., de Robin, 1869, p. 560.)

These facts show, then, the enormous influence which the blood exercises upon the vegetative phenomena of the life of the nerve-cells, and to what an extent their individual dynamic activity, and consequently the life of the whole system, depends upon it.

It is the blood that carries everywhere with its uninterrupted currents the vivifying stimulation which causes the cells to feel, to become erect, and to associate for co-ordinated actions. In the purely sensitive regions, where the phenomena of conscious personality are incessantly in process of evolution, it keeps them constantly awake, and thus sustains in us the conscious idea which we possess of the external world. In the motor regions it enables the nervous elements to accumulate, as in condensers, a store of nervous influence destined to pass into the dynamic condition as soon as a call is made upon them. It is everywhere present, flowing everywhere, and evoking the specific innervation of each of the cell-territories which it animates and bedews, thus enabling them to renew their latent energies.

When once provided with the necessary elements of nutrition, the cerebral cell becomes capable of entering into action, and performing the dynamic function for which it is designed. This new phase under which it reveals itself is characterized:—

1. By an acceleration of the blood-currents in the functioning regions.

2. By a local development of heat in these regions.

1.—If it be incontestably demonstrated what an important influence the regularity of circulatory phenomena has in evoking the activity of nerve-cells, it is on the other hand very curious to note what an influence the activity of these same cells may have in return on the vascular irrigation designed to provide for their nutrition as well as their expenditure.

It is not, indeed, without a certain astonishment that we observe that if, on the one hand, the nerve-cells play a passive part with regard to the circulation which feeds them—if they are in subjection to it, and are veritably its tributaries; by an inverse phenomenon, from the moment they become active their position changes, and, ceasing to be subject as they were, they in turn become dominant. From the very fact that they are working—that there in certain isolated spaces they develop a state of nervous erethism—they at the same time determine *hic et nunc* a concomitant influx—they make an appeal to the blood, and even turn to their own profit the irrigation of certain neighbouring regions.*

Thus the brain, as regards the phenomena of circulation, is at the same time active and passive; it is of necessity subject to their influence, and cannot, on pain of cessation from all work, refuse their aid; and yet, at a given moment, it reacts, solicits them, makes appeals to them, and thus unconsciously directs the vasomotor actions designed to maintain the integrity of its vital energy.

Thus from this double influence of the phenomena of the circulation on those of cerebral activity, and those of cerebral activity on the acceleration of the

* We may perhaps attribute to an accidental derivation of blood towards a circumscribed portion of the brain that is in a state of erethism, and the consequent draining of the circumjacent regions, certain phenomena of cerebral life in which, under the impress of a strong preoccupation, a concentration of the mind upon a single point, we momentarily lose the notion of the surrounding medium, and cease to perceive what is passing around us.

flow of blood, a vicious physiological circle results, calculated to have an inevitable influence upon the infinite series of regular cerebral operations, as well as upon the progressive evolution of pathological phenomena, which mostly are but exaggerations of the normal actions of the organism.

Every one knows how fatal chronic lesions of the capillary plexuses are to the delicate substance of the cerebral cells—how the plastic exudations which proceed from the vessels, the fibro-albuminous deposits which become infiltrated into the tissue and interstices of the cells, become like so many foreign bodies hostile to life, and injurious to the physiological medium whence they draw the elements of their normal constitution.

Every one knows, further, how moral causes—too energetic work, which exceeds the amount of the reserved nerve-force — prolonged vigils, which do not permit the recuperation of lost materials—preoccupations concerning a single subject, which induce a condition of chronic congestion within certain circumscribed limits—are so many morbid modes of excitement which maintain a permanent condition of local erethism, and thus indirectly become the causes of those repeated affluxes of blood which are so inevitably followed by exudations of all kinds and persistent new-formations (the lesions of general paralysis).

Hence that preponderant influence which the whole series of moral affections exercises upon the genesis of mental maladies. Whether they be derived from an intellectual excitement prolonged beyond physiological limits, or result from profound disturbances occurring in the emotional sphere of the sensorium, in consequence of trouble, disappointment, misfortunes of all kinds, the minute mechanism of their advent is always fundamentally the same. It is by the physiological channel they introduce themselves into the organism; it is under the form of regular excitations—shocks propagated along the normal processes of cerebral life—that they implant, develop, and perpetuate themselves; and the incurable disorders they leave behind them are but the indirect effects of disturbances of nutrition in the nervous plexuses, proceeding from this single source, the afflux of blood too frequently provoked.*

Sleep.—By an inverse phenomenon, if the cerebral cell, from the very fact that it is in its period of erethism, its working period, becomes the occasion of a call upon the blood destined for its activity, this curious fact occurs, that so soon as this activity begins to slacken, so soon as fatigue announces itself, and its histological sensibility is exhausted by the action of external impressions, the vascular irrigation is modified simultaneously. It follows step by step the decreasing phase of the dynamic activity of the cells that depend upon it, and at the same time that the brain becomes weary, and that the sum of its functional energies diminishes, the mass of blood which flows to it becomes less, the capillaries

---

* Calmeil thus expressed the same thought. "All the so-called moral influences, whether they betray themselves by the persistence of annoyances or regrets, or take the forms of jealousy, hatred, or ambitious disappointments, may concur to produce a morbid accumulation of blood in the encephalic capillaries. (Calmeil, " Maladies inflammatoires du cerveau," vol. i. p. 5.)

See also Forbes Winslow, on softening of the brain occurring from anxiety and forced exercise of the organ, and consisting in feebleness of mind. ("Annales médico-psychol," 1850, vol. ii. p. 711.)

are less gorged with blood, and the cerebral tissue insensibly becomes exsanguine.

This is that new state of cerebral ischæmia, opposed to the phase of congestive activity, which as an alternative fact of the general order that exists in the brain of all living beings, inevitably reveals itself whenever their cerebral cells, having exhausted their accumulated nervous forces, become fatigued by exercise and fall into the physiological collapse of sleep.*

Where the life of the nervous elements is stilled, a stillness also takes place in the most minute currents of the circulation, and these two phenomena, which act and react on one another in the ascending phases of activity, similarly affect each other in its descending phases. When the vital movement becomes slack, and histological sensibility dull, the demand upon the blood is less imperious.

2.—If, from a chemical point of view, the phenomena of cerebral activity are characterized and gauged in a precise manner by the real loss of brain-substance, and the passage of a certain quantity of phosphorized

---

* The condition of comparative anæmia in the brain during sleep has been directly proved by different observers; thus Caldwell, in the case of a wound in the head, with loss of substance in the bones of the *cranium*, observed that when the patient was plunged in deep and peaceful sleep, the brain remained almost immovable in its envelope, but that when he was dreaming it increased in volume, and when the dream was vivid it protruded through the opening. Blumenbach in an analogous case, similarly remarked that the brain subsided during sleep, and that waking was accompanied by a more or less considerable afflux of blood, and an augmentation of volume. ("Archives générales de Médicine," vol. i. 637.) Durham also has instituted direct experiments to prove that during sleep the brain becomes anæmic. ("Guy's Hospital Reports," 1860, vol. vi. p. 149.) See confirmatory experiments by Claude Bernard. "Leçons sur les anesthétiques," p. 117. Paris, J. B. Baillère, 1875.

matters in the urine, from a physical point of view they present characters which are no less significant, and no less important to recognize.

The authors who have already occupied themselves with the question, as to what appreciable physical modifications are presented by the brain-substance while in activity, have noted in a precise manner that this inward labour reveals itself by sensible signs, in the form of a more intense disengagement of heat; and that the brain, like a muscle in action, manifests its dynamic power by a local increase of heat, appreciable by the instruments of the physical laboratory.

Thus, Lombard (of Boston), who was the first to institute experiments in this direction, arrived at the following results, by means of very exact thermo-electric apparatuses:—

"In the condition of cerebral repose," he says, "during wakefulness, the temperature of the head varies very rapidly. The variations are very slight, not attaining $\frac{1}{100}$th of a degree centigrade, but they are not the less worthy of attention, for this reason—that they are confined to the head.

"The variations of temperature appear to be connected with different degrees of cerebral activity. During active brain-work it never exceeds $\frac{1}{20}$th of a degree centigrade.

"Every cause that attracts the attention—a noise, or the sight of an object or a person—produces elevation of temperature.

"An elevation of temperature also occurs under the influence of an emotion, or during an interesting reading aloud.

"This elevation of temperature is especially well-marked in the region of the occiput."

These experiments, as we see, apply only to the appreciation of temperature externally estimated, on the skin of the cranium. The brain was not directly investigated.

Schiff has supplied this omission, he has entered the cranium, and by means of thermoscopic instruments of extreme sensibility has succeeded in directly examining the cerebral substance at the moment when it came in contact with external excitations, and thus determining what degree of elevation of temperature the brain is susceptible of attaining in its operations.

This ingenious physiologist has therefore succeeded in defining not only what regions of the cerebral cortex are isolatedly called into play by such or such kinds of sensorial impressions, and demonstrating experimentally that there are isolated circumscribed spots reserved for such or such kinds of sensorial impressions (as has already been described on the authority of anatomy); but also that the arrival of these impressions resolves itself into a local development of heat in the special area where it disseminates itself; and that the heat thus developed is a dynamic phenomenon independent of the circulatory activity, a true vital reaction of the sensorium—that, in a word, it is the direct result of the participation of the psychic element on the arrival of the sensorial excitation.

"Thus," he says, "the psychical activity, independently of the sensorial impressions which call it into play, is connected with a production of heat in the nervous centres, a greater amount of heat than that which simple sensorial impressions engender. This conclusion is justified by the decrease of the calorific effect of a strong and always identical sensorial impression, which animals have been made to experience many times in succession. Let us take a pullet," he adds, "whose sight or hearing we assail by appropriate means. The first impression which the unprepared animal receives will excite in it more intense psychical reflex actions than the succeeding excitations of the same nature, since it insensibly becomes habituated to them." Thus, by eliminating gradually the part played by psychical action in sensorial absorption, he arrives at an estimate of the heat evoked by the arrival of simple sensorial impressions, and that which proceeds from the direct participation of the psychical activity at the beginning of the experiment.*

We thus understand, after this series of experiments, how prolonged efforts of the mind, and moral emotions of all sorts, from the very fact of their awaking the activity of the *sensorium*, are calculated to have an immediate effect upon the essential phenomena of the nutrition of the brain.

They show us, indeed, on the one hand, that sustained intellectual work is accompanied by a loss of phosphorized substance on the part of the cerebral cell in vibration ; that it uses it up like an ignited pile which is burning away its own essential constituents ;† and that,

* Schiff, l.c., "Archives de Physiologie," 1870, p. 451.
† Louyer-Villermay cites the example of a celebrated lawyer who lost his memory in consequence of too long-continued intellectual work; and Moreau de la Sarthe reports a similar case which occurred in a German savant, after an intense concentration of mind. We also know of a great number of musicians who have become deaf by the immoderate exercise of the organ of hearing. "Journal d'hygiène," 1875. (De la surdité chez les Musiciens, par Dr. Prat.)

on the other hand, all moral emotion perceived through the *sensorium*, all effective participation of this same *sensorium* in an excitation from the external world, becomes at the same time the occasion of a local development of heat.

These facts are destined to have a direct effect upon our knowledge of the essential conditions of the integrity of the cerebral functions, and to formulate absolute hygienic principles with regard to them.

It stands to reason, indeed, that if the cerebral cell expend its reserve material during its diurnal activity, it is absolutely necessary, to enable it to continue alive and in health, that it shall repose and sleep regularly. Sleep is to the brain, what needful repose is to our fatigued limbs, the necessary condition of its health. Every one knows, indeed, how great is the number of individuals who have sown the seeds of a cerebral disease by a prolonged infraction of these simple laws of hygiene, and who through reiterated vigils and exaggerated expenditures of activity, have thus passed the physiological limit of the resources at their disposal, and incurred expenditure above their receipts.

On the other hand this development of heat, which is produced in certain circumscribed localities of the brain when an emotion or sensorial impression is reverberating through the plexuses of the *sensorium*, further shows us with what circumspection we should manage this kind of excitation in individuals whose brain is in a painful condition, either from a recent congestion, or from former congestions grafted one upon another.

We all know from more or less personal experience, that when we have a headache, and our *sensorium* is in a state of hyperæsthesia, the smallest noises, the slightest external incidents, produce in us painful shocks, and that the absolute incapacity for work is most painful.

All doctors know how often, in persons excited by the occurrence of repeated cerebral congestions, paralytics, maniacs, and even patients with certain forms of melancholia, the unexpected calling up of an old emotion, the sight of a relative, may have a sad effect upon their cerebral condition. We see, indeed, their faces redden and grow pale, and very often the effect of an emotion inopportunely provoked, is but the prelude to the return of more and more serious congestive accidents.

PART II.

GENERAL PROPERTIES OF THE NERVOUS ELEMENTS.

THE phenomena of the life of the nervous centres, spite of their apparent complexity, are nevertheless regulated by laws which are in general simple—common principles, which indisputably give them an air of near relationship. These common principles are, moreover, themselves reducible to elementary vital properties, which form the basis of each of them in particular, and constitute, in a manner, the simple primordial principles which we constantly find underlying every combination of nervous activity, however complicated it may be. These fundamental properties, which thus serve as elementary materials for every dynamic action of the system, may at the present day be thus epitomized, under three principal heads:—

1. *Sensibility*, by which the nerve-cells feel excitation from without, and react in consequence, by virtue of the excitement of their natural affinities.

2. *Organic Phosphorescence*, by which the nervous elements, like bodies which have received the vibrations of light, preserve for a prolonged period traces of the

G

excitations which have in the first place set them in action, thus storing up within themselves phosphorescent traces, which are records of the received impressions.

3. *Automatism*, which expresses the spontaneous reactions of the living cell, which sets itself in motion of its own accord (*motu proprio*), and in an unconscious and automatic manner expresses the different states of its sensibility thrown into agitation.

It is the history of these various general properties of the nervous elements that we are now about to study in due course, in the physiological part of this work. These properties once defined and known, we shall attack the study of the different combinations to which they adapt themselves by combining one with another; and thus, proceeding from the known to the unknown, from the simple to the composite, we shall advance with better ascertained points of support into that domain, so complex, and at the same time so rich in interesting prospects—that of cerebral activity proper.

BOOK I.

SENSIBILITY OF THE NERVOUS ELEMENTS.

CHAPTER I.

GRADUATION AND GENEALOGY OF THE PHENOMENA OF SENSIBILITY.

SENSIBILITY is that fundamental property which characterizes the life of cells. It is by means of it that the living cells come into contact with the medium that surrounds them, and that they react *motu proprio*, by virtue of their natural affinities which are thrown into agitation, and exhibit a desire for the excitations which gratify them, and a repulsion for those that are unpleasant to them. Attraction for agreeable and repulsion for disagreeable things are the indispensable corollaries of every organism fitted for life, and apparently the elementary manifestation of all sensibility.

Sensibility, which is, perhaps, itself, in the organic world, only the transformation of those blind forces, which attract among themselves the crystalline molecules of the inorganic world, and group them according to their proper affinities ; this phenomenon, sensibility, begins to appear, in its most simple forms, with the first rudiments of life.

It is in the unicellular organisms of the vegetable kingdom that it first embodies itself and reveals itself in its own shape; and here it shows itself as a property of tissue, very distinctly connected with the very substance of the amorphous protoplasm of which it is the endowment, under the form of vague diffuse contractility, no special element being reserved for it, and no nerve-cells being as yet extant.*

Little by little, as the living cells group together and form more dense agglomerations, the phenomena of sensibility become more distinctly evident, and soon we find them provided with special apparatuses designed to serve them as a support, and to condense and perfect their modes of activity; while in the superior animals they become more and more highly endowed, to arrive at man as the last term of their long evolution, and produce those phenomena so rich, so varied, so delicate, defined *in concreto* under the name of the moral sense.

In this chapter we shall follow the process of the evolution of sensibility, from the most elementary phases under which it shows itself at its point of origin, to the moment of its most complete expansion in man.

Sensibility, we may say, in its most simple revelations in unicellular organisms, at first appears in a vague and undetermined form. It reveals itself by that essential tendency which these protorganisms have, to seize upon substances which gratify their natural affini-

* Naturalists have made known to us beings of an organization so simple that their entire body is formed of but one cell. Their whole development, their whole existence, is shut up within limits thus strict. We may mention the *gregarines* in particular. (Frey, " Histologie et Histochemie," p. 74.)

ties and avoid such as are inimical to them. It regulates and governs the continuity of the purely trophic phenomena of the life of cells.*

In vegetables the phenomena of sensibility have already taken more distinctly marked forms. Their cycle is no longer restricted to the local operations of rough and ready assimilation and disassimilation.

Vegetable cells, even when agglomerated in but small groups, have become sensitive and impressionable by external agents. Calorific and luminous impressions produce a certain effect upon them, and if this effect be grateful to certain natural affinities, we may see them gradually inclining in the direction from whence these excitations come. They turn automatically towards the sun, awake with him when he appears, sleep when he has disappeared, and, in a word, present that series of unconscious and graduated movements by virtue of which they tend towards the realization of their latent satisfactions.†

Botanists have already described those curious phenomena of vegetable sensibility by virtue of which we see the petals of certain flowers fold up at night and unfold in the day time; the stamens of the barberry,

---

* The *gregarines*, which are met with in troops as living parasites in the alimentary canal of insects and other animals, are not only destitute of a mouth, but even of vibratile *cilia*. They are simple cells with apparent nuclei. (Hartmann, "Conscience des plantes." "Revue Scientifique," July 1873, p. 623.)

† Plants which catch insects are sensitive to the touch; climbing plants discern points of support. The leaf of the vine feels the light, towards which it strives to turn the right side, and every flower feels it, and strives to bend its head towards it. The *mimosa* feels and reacts. It is the essence of every motion that it shall be preceded by sensibility. (Hartmann, *loc. cit.*, p. 625.)

under the excitement of a light touch apply themselves to the pistil; the flowers of the water-lily hide themselves at the bottom of the water while they wait for the day. It is even more astonishing to see what happens with sensitive plants, and to observe how that curious vegetable, *mimosa pudica*, presents in itself all the most delicate manifestations of the impressionability of living beings.*

Like an animal, it feels and reacts on the contact of the lightest touch; feels inequalities of temperature;† is influenced and struck with anæsthesia by the inhalation of chloroform; like an animal, moreover, its sensitive unity forms a complete whole; its leaflets and rootlets are united in such an intimate *consensus* that if its rootlets be subjected to the action of any irritant, its leaflets are affected at the same time, and sympathize painfully with their sister cells of the lower regions which have been thrown into agitation; just as we see that sensibility when developed in any region of an animal whatever, has a generalized reaction all over the organism.

* Marked movements are performed every evening by vegetables with composite leaves, like the *cytisus* or *robinia pseud-acacia*. We see these plants make their preparations for night every evening—some simply fold their leaves, others, with more foresight, prudently enclose their flowers. The great lotuses of the Nile, and the water-lilies of our own lakes, draw down their carefully closed corollas to the bottom of their waters; and the sun must have come next day to illumine the earth before the chilly and sleeping plant consents to open its petals.

The sleep of plants is related to the greater or lesser intensity of the light with which they are surrounded; and, what is more conclusive, plants which have been strongly illuminated at night, while they are in obscurity during the day, have changed their habits so as to sleep in the day and wake at night. (Edmond Grimard, "De la sensibilité végétale." "Revue des Deux Mondes," 1868, p. 379.)

† Grimard, *loc. cit.*, p. 385.

THE PHENOMENA OF SENSIBILITY. 87

In the animal kingdom sensibility reveals itself in its origin by phenomena exactly comparable with those which we have just sketched.

There, in the form of amœboid movements of the white corpuscles and ciliated cells, and contractility of the protoplasmic cells,* it shows itself to us under the appearance of purely histological sensibility, and not as yet in the shape of sensibility belonging to a living autonomous individuality.

In the protozoa, rhizopods, and certain polyps, it becomes more and more distinct, and by the very complex operations through which it manifests itself, we perceive how well these protorganisms of the animal kingdom are provided with active and vital energy, and how distinctly general sensibility is inherent in them and combined with their substance.

In these elementary forms of animal life, the phenomena of sensibility are first united with an organized tissue. They are divided among as many cells as the individual contains; and they exist in a vague and diffuse manner, without there being as yet a special system of anatomical elements, designed to serve them as an appropriate receptacle.

Soon, as we ascend in the series of beings, new factors are added to the preceding; the phenomena become complicated as they grow more perfect, and we then see that in proportion as animal organisms develop themselves, and their agglomerations of cells become more numerous, there takes place among them, as it were, a natural selection of the physiological work to be performed. Some are gifted with such or such specific

* Wund, "Physiologie," p. 83.

aptitudes, and appropriate such or such a function, while others, gifted with such or such a different aptitude, reserve themselves for such or such another. For its better performance there is a division of labour.

This natural division of the living forces of the living individual, which are thus distributed among the different departments of its substance, constitutes the first outline of the nervous system.

It soon appears, like an organ of perfectionment implanted in the organism. It is henceforward the grand dispenser of sensibility in general, and is designed to collect, to *drain* all the scattered forms of sensibility, to regulate their course, to condense them in its own reservoirs, to purify them by the participation of its substance, to make them leap forth in the form of motor excitations, or to transform them, like the perfected products of its own industry, into subtle and quintessential materials, destined to co-operate in the most subtle phenomena of psycho-intellectual life.

Humble in its origin, the nervous system, as F. Leydig has pointed out, makes its first appearance in the midst of the living tissues in the form of three or four cells, independent one of another.* One step further, and the cells are united within a common envelope, a first nervous ganglion being thus constituted. Little by little the work of evolution completes itself; ganglion is united to ganglion; these soon dispose themselves in the form of two lateral rows, which emit, right and left, radicles which plunge into the surrounding tissues, and soon these two lateral chains, approaching, become fused together, and thus constitute a central unity,

* Claude Bernard, "Système nerveux," vol. i. p. 506.

THE PHENOMENA OF SENSIBILITY. 89

or axis, around which all the nervous radii emerging from the peripheral regions converge. At the same time, a superior ganglion, destined to be the brain, is developed, and uniting itself to the axis, becomes in a manner the crowning of the edifice thus successfully perfected.

From this moment the nervous system is constituted as a central force destined to condense in its plexuses sensitive excitations, in order to transform them by its own metabolic action into co-ordinated motor reactions. From this moment the living forces of the organism are duly subordinated and distributed in a methodic fashion; the physiological task is regularly divided; one group of elements is connected with sensibility, one centre with motor-power, and another with the functions of organic life.

Sensation is henceforward neatly isolated in special regions of the system, neatly collected in particular organs, and from the very fact that it is attracted, like an electric fluid, by means of nervous conductors, from the peripheral regions towards the central, it becomes a disposable mobile force, transmissible to a distance like dynamic electricity.

Once concentrated in the central regions of the system, it thus represents, with all the diverse elements of which it is composed, a true synthesis of all the partial sensibilities of the living being, and the true generating element of its living and feeling unity.

The phenomena of sensation in the superior animals are not, then, simple phenomena, constituted by the mere reaction of a tissue in the presence of external excitations; they are the complex subordinated operations of the nervous activity which require the participation of a great many organs successively brought into play, in order to arrive at their complete evolution. We shall now study these different conditions in succession.

CHAPTER II.

EVOLUTION OF THE PROCESS OF SENSIBILITY, THROUGH THE MECHANISM OF THE NERVOUS SYSTEM—UNCONSCIOUS SENSIBILITY — CONSCIOUS SENSIBILITY (SENSATION).

THE nervous system being constituted, as we have just explained, by a central axis, plunging by its lateral roots into the surrounding tissues, and crowned at its superior extremity by a central ganglion, the brain, gifted with its special activity, we shall now see how the phenomena of sensibility, existing *per se* as fundamental histological properties, behave in presence of the machinery which the nervous system places at their disposal; how they become incorporated with it; how, arriving in the form of centripetal excitation, they become refracted in the plexuses, reappearing as a centrifugal reaction, through the peculiar influence of the new *media* they have put in requisition; and how at last, in the most elevated regions of their journey, they come to play a primary part in the evolution of the essential phenomena of psycho-intellectual activity.

In taking their departure from the peripheral regions of the nervous system, which physiologically represent the frontiers of the organism, sensitive impressions, wherever they may have originated, once implanted in their tissues in the form of vibratory agitations, follow their natural channels towards the central regions.

Some are extinguished in certain interposed ganglionic masses; others advance further, become dispersed in the grey regions of the cord and transformed, either instantaneously or in a more or less gradual manner, into excito-motor reactions—these being the phenomena of unconscious sensibility.

Others, finally, endowed with an altogether special vitality, pursue their course, converge, mount up to the *sensorium* and come into contact with the psycho-intellectual operations for which they provide the indispensable food—these being the phenomena of conscious sensibility, or sensation, to which they give birth.

We shall successively pass in review the mode of genesis and distribution of these two special groups of sensitive contingents.

Unconscious Sensibility.—Unconscious sensitive excitations are derived from two orders of peripheral plexuses:—

1. From the plexus of vegetative life of the sympathetic.

2. From the plexus of general and special sensibility.

These latter originate in common with the excitations destined to ascend to the *sensorium*; but they are extinguished on the way, and are destined to produce reflex actions (automatic actions) in the interior of the plexuses of the spinal cord.

1. Sensitive excitations radiating from the plexus of vegetative life, if we take them from their origin, only expand within a limited radius. They follow the threads of the sympathetic, which are distributed *ad*

infinitum throughout the organism, and only manifest their presence by vaso-motor phenomena, capable of modifying, in a more or less direct manner, certain branches of local circulation.

This special order of sensitive impressions is condensed in special ganglionic masses, which represent small local centres, and are the primitive types of the first traces of a nervous system in the lower species.

Sometimes they are capable of radiating to a distance, and thus traversing several ganglionic masses and vibrating even as far as the grey plexus of the spinal cord, of which they thus provoke the secondary activity. Thus it is that the sensibility of the intestinal mucous membrane excites the secretion of the juices destined to co-operate in digestion; that the sensibility of the uterus laden with the product of conception leads to development of the breasts; that in abnormal conditions certain abnormal sympathies are developed, so that we see, for instance, the irritation of the urethral mucous membrane exercise an influence upon certain articular surfaces; and that the irritation of certain peripheral nerves leads to the sudden occurrence of tetanic phenomena and of certain, so called, *reflex* convulsions.

Sometimes also, when certain peripheral regions in which they originate are intensely affected, and have risen to the pitch of pain, the excitations of sensibility become capable of an action more penetrating still, and even of reaching the *sensorium*, where they are perceived, and whither they carry, as it were, the cry of some organ of vegetative life shaken in its essential constitution.

In general we may say, that in the normal state the impressions of vegetative life are quite silent and unperceived by the *sensorium*. The wheels of the inner life of the human machine move without noise. Few persons except medical men, are aware that they possess a heart provided with auricles and ventricles, which contract alternately a great number of times a minute; a stomach which secretes a juice destined to dissolve the azotised elements of the food; a pancreas designed to act by means of its secretion upon the fatty elements; intestinal fibres which contract alternately and force along the alimentary bolus, &c. All these phenomena take place without our knowledge, without our having the slightest notion of them, and, strange to say, those facts in which we are most vitally interested we know least about!

But is this really the case, and are we authorized to think that the different forms of sensibility, which are in activity in the inmost recesses of our tissues, really exist without having a sort of obscure influence upon our *sensorium*, analogous in this respect to those obscure rays of the spectrum which our eyes do not behold, and which yet have so real and indubitable an existence?

This does not seem probable to me; for if we think how instantaneously a visceral pain is developed, with what clearness this pain appears when a calculus is fixed in the *ductus choledochus*, or when a foreign body is introduced into the stomach or intestines, where it instantly produces painful contraction, we cannot help thinking that there are always open roads between the *sensorium* and the regions of vegetative life; that there is, in some manner, an incessant relation between these two poles of sensibility; and we must recognize the fact that there is, in the normal state, a constant though unconscious afflux of the partial sensibilities of the organism which converge towards the centres, and that they die away there in silence without making any impression, yet bringing an unconscious notion of all that passes in the periphery of the nervous system. We see every day substances with which we are constantly in contact, from habit pass us by unperceived, leaving in the *sensorium* only an unseized impression, like that produced by the atmospheric air on the respiratory tract. Water and bread, which are so frequently in contact with our digestive mucous membranes, furnish us with but obtuse impressions, which yet are consciously perceived.

It is then probable that if the sympathetic nerves of vegetative life, starting from the peripheral regions, form a continuous network, of which the converging meshes more and more nearly approach the central regions, the histological sensibility which they abstract from the different cell-territories, amidst which they originate, follows the same natural channels; and that this is led up to reverberate within the *sensorium*, in a disconnected obscure manner it is true, yet, nevertheless, really and permanently.

We cannot fail to recognize in this afflux of all the diffuse sensibilities of the organism, each coming to bring to the *sensorium* its sensitive note, that series of generating elements which are designed to implant themselves there and develop in us that essential notion of our vital being, which makes us feel ourselves live in all our organic molecules. It is in itself nothing but the unconscious notion of all the partial sensibilities of the organism, concentrated in this grand common reservoir.

2. Unconscious excito-motor impressions arise, with their sister conscious impressions, in the terminal expansions of all the sensorial and sensitive nerves.

Mingled with their fellows they enter the converging channels which are open to them, and advance together with them towards the central regions of the spinal axis, having, however, first traversed the chain of the rachidian ganglions.

Arrived at the grey plexuses of the spinal axis, they become diffused in their meshes, excite the activity of the posterior grey regions (which represent, as it were, a great common *sensorium* of unconscious life, for this order of radiations), and pass out in centrifugal currents, in the form of co-ordinated motor reactions, which thus represent the last phase of a process originating in the purely sensitive regions.

The unconscious excito-motor sensibility, transformed by the action of the cells belonging to the automatic *sensorium*, by this very circumstance acquires new properties.

It is stored up, seized upon, and condensed on the spot in the tissue of the organs that receive it, thus becoming in this new form, like a projectile rammed home in a fire-arm, capable of being transmitted to a distance along the centrifugal conductors radiating from the spinal cord, veritable *reophores* designed to favour its dissemination and transport it to a long distance, even into the most distant and eccentric cell-territories.

Thus it directs, in the form of unconscious optic

EVOLUTION OF SENSIBILITY. 97

excitations, the different movements of rotation of the ocular globes, the play of the pupil, the accommodation of the sight to different distances; excites in the sphere of auditory phenomena the unconscious movements of the chain of little bones, to graduate the alternate tension and relaxation of the tympanic membrane; co-operates so powerfully in the complex and varied movements of mastication and deglutition; presides over the succession of the acts of erection and ejaculation; and, in a word, in different forms, without the intervention of the *sensorium*, always present, always active, assists in the perfecting of the sense to which it is attached, favours its direction towards an object, governs the play of its mechanism, so as to obtain the maximum of sensorial impression, and thus becomes the indispensable adjunct of conscious impressions.

It is still this unconscious excito-motor sensibility that underlies the different processes of the respiratory phenomena during the whole term of our lives, from our first inspiration to our last sigh.

It maintains the play of the motor ganglions of the *medulla oblongata*, those central *foci* of innervation, whence the inspiratory and cardiac muscles draw their unceasing principle of activity. It expends itself at every instant, day and night, in the continual activity of the mysterious laboratories of organic life. It moreover plays an all-important part in the varied series of our movements of progression, in all those of bodily exercise, in the methodical motor actions that we insensibly bring to perfection by practice and sustained attention—such as those of the hand in drawing or writing—actions which though at first conducted with the conscious participation of the *sensorium*, insensibly come to be executed under the sole direction of the excitations of unconscious sensibility.

Thus, then, the phenomena of automatic life, under whatever form they present themselves, occur of themselves and by virtue of the intra-spinal transformation of an incident excitation of reflex sensibility into motor reaction; and this without the *sensorium* coming into play, without the intervention of conscious sensibility, simply as a return effect of the calling into activity of a process of unconscious sensibility.

But although the phenomena apparently take place thus, being evolved without the effective participation of the *sensorium*, it must not be concluded that no fraction whatever of these excitations is radiated towards it and extinguished in it, somewhat in the manner of obscure rays.

It is very probable that what we have seen to occur as regards the impressions of purely vegetative life may occur as regards this special order of excitations, there being probably an obscure radiation of these latter impressions which extends to the *sensorium*, and thus transmits to it the vague and unconscious notion of the activity of such or such a portion of our muscular system.

If the *sensorium* indeed be not directly active in the infinite series of motor acts that we accomplish automatically, it nevertheless does not remain a complete stranger to the operations which take place within the organism. If it does not interfere directly to regulate the play of such or such an organ, to move, for instance, the crico-arytenoid muscle in a methodic manner for

EVOLUTION OF SENSIBILITY. 99

the production of such or such a laryngeal sound, or the accomplishment of such or such an act of digital dexterity; if the conscious personality cannot discern who are the workmen at work, it has at all events an exact notion of the operation in evolution, knows if the work be accomplished, and the requisite muscular exertion made. We do not feel our muscles in a clear and precise manner when they are in a state of repose; but when they are in activity, this new condition into which they are thrown develops in the *sensorium* a new mode of existence, so that the unconscious excito-motor sensibility in the dynamic state indirectly strikes upon the *sensorium*, and thus becomes a new element destined to become absorbed in its plexuses.

Conscious Sensibility (Sensation).—The sensitive excitations destined to become conscious and enter into relation with the phenomena of psycho-intellectual activity, are collected, with their excito-motor fellows in the peripheral plexuses, which serve as a region of emission for both. Starting from this, and taken up by means of the converging fibres, they pass on towards the central regions of the axis, are concentrated in the isolated ganglions of the optic thalamus, and are afterwards radiated, as we have already seen, into the different regions of the cortical periphery. (Fig. 6.—9. 4. 14.)

The phenomena of conscious sensibility (or sensation) have then as their point of origin, and first halting-place, the peripheral regions of the nervous system. It is by the terminal nervous expansions spread out into a network, open, in a manner, to all that comes to impress it, that the external world penetrates and becomes incarnate in us. And for this a special faculty for

H 2

receptivity and impressionability in the nervous element thus impressed is in the first place necessary, as a fundamental and indispensable condition of the phenomenon.

In a word, it is necessary that at the moment the sensorial network receives the vibratory excitation, it shall directly participate in the act which takes place within it. It must become active, acquiesce—become in a manner erect; and must, by a species of vital assimilation, convert the purely physical into a physiological excitation, the luminous vibration, for instance, into a nervous one.

This is the fundamental act of which we shall speak again subsequently, and which is the first link of that chain of sensitive phenomena which is evolved throughout the nervous system.*

It is, in fact, a vulgar truth which reveals itself to simple observation. Every one knows that the simple presence of a physical excitation of a sensorial organ is insufficient to produce a conscious impression, and that an active participation of the sensorial cell in the vibratory movement communicated to it is necessary. Open the eye of a sleeping man—the luminous rays fall in vain upon the retina. It requires a certain time before the nervous cells are wakened up and enter into harmony with the luminous vibrations which

* This phenomenon has been perfectly described by Mathias Duval. "When the retina is excited," he says, "perception is not immediate, it is retarded for a very short period; this retardation is due to the fact that it requires a certain time for the transformation of the luminous into a nervous movement to take place. Then this latter movement requires an interval, exceedingly short indeed, to be propagated along the optic nerve to the cerebral centres; and finally, the centres of perception themselves are not immediately thrown into agitation. This retardation occupies one-fiftieth to one-thirtieth of a second." (Mathias Duval, "Thèse d'agrégation," 1873, p. 132.)

excite them. Pinch the skin of a man in profound sleep, cry into his ear under the same conditions. There is the same apathy, the same default of reaction. The purely physical excitation will gradually become deadened if there come not in its train a purely vital phenomenon of sensation, which is developed, by a sort of active prehension of the physical food which is offered to the impressed cell.*

We see, then, judging by what takes place here in this first phase of nervous activity, that the sensitive plexuses of our whole organism are all either isolatedly or simultaneously thrown into vibration, according to their various tonalities. They thus become like vast vibratory surfaces, of which the oscillations, registered as they arrive, are incessantly transmitted to the other parts of the system, and felt in the *sensorium* in a corresponding manner. It is a continuous, regular, imperative work, which is accomplished every moment, from the peripheral to the central regions of the system, and this uninterrupted appeal from the external world is so necessary, so much the obligatory condition of all cerebral activity, that the latter ceases at once when its means of alimentation from without are cut off (loss of consciousness, sleep, lethargy), just as we see the phenomena of hæmatosis cease, when the atmospheric air suddenly ceases to enter the recesses of the respiratory channels.

* The participation of the sensorial element in the external perturbation is itself only a species of fugitive phenomenon, having a transient duration. When the duration of the impression is too prolonged, transgressing physiological limits, it brings on a period of fatigue of the receptive element, and ceases to produce any effect. Thus the sensibility of the retina is rapidly blunted. When, for instance, we look for a long time at a white spot on a black surface, and afterwards look at a white surface, we imagine that we see a black spot upon this, the retina having become insensible to white.

CHAPTER III.

INTRA-CEREBRAL PROPAGATION OF THE PROCESSES OF SENSIBILITY.

SENSITIVE impressions pursuing their course are, as we have already seen, condensed in the masses of grey matter in the optic thalami.

These masses of grey matter represent, then, in the general economy of the nervous system, a species of point of convergence, or cross-roads, and the penultimate halting-place where impressions from the external world are united before being radiated towards the peripheral cortical regions.

But as regards these different kinds of sensitive elements which come flowing towards the grey ganglions of the optic thalami, these latter, which receive them into their mass, give them each an isolated territory—so that that division of labour of which we have already seen an example in the progressive evolution of the nervous system, here seems to receive a new confirmation, since we see the phenomena of sensibility divided, like white light, into isolated fascicles, each fascicle having a special receptive apparatus reserved for itself exclusively.

Thus purely sensitive impressions have a central ganglion where they are isolatedly condensed (Fig. 6—9.); it is the same for the optic, olfactory, and acoustic impressions and finally the excitations of vegetative life also find a cell-territory specially appropriated to their reception,—so that as the processes of sensibility become perfected, as they penetrate more deeply into the interior of the nervous system we find them splitting up, dividing into elementary fascicles, each gifted with dissimilar specific properties, and yet united among themselves by the common bonds of their origin and evolution.

After radiating through the cerebral white fibres, into the different departments of the cortical substance, the same phenomena of division of labour again occur, and we may directly observe that the regions in which the dissemination of auditory impressions takes place are different from those where that of the olfactory, visual, etc., takes place. So that each isolated region of the brain has also to work and develop its specific energies in isolation. (*See* Fig. 6—4. 9. 14. and Fig. 5—7. 8. 10.)

When the sensitive excitations, whatever they may be, have been launched into the midst of the plexuses of the cortical layer, they find there also sensitive nervous apparatuses prepared to receive and absorb them, and thus co-operate in the various processes in evolution.

We have indeed already studied the remarkable disposition of the cells of the cortex (*see* Fig. 1.), which are arranged in isolated zones, stratified like the layers of the crust of the earth, and thus constitute a continuous network of which all the organically connected molecules are arranged so as to vibrate in unison, and to propagate the nervous undulations, either vertically or laterally

On the other hand, those myriads of nerve-cells, agglomerated into a continuous whole in the sub-meningeal regions of the cortical substance, are themselves essentially sensitive. They are living, impressionable, and gifted in the highest degree with that vitality which characterizes the nervous elements: and accordingly, when the perturbation from the external world, transformed by the metabolic action of the optic thalami, comes to reverberate within them, they are perturbed in their turn, and are in a manner thrown into a condition of erethism, just as the peripheral plexuses were when first agitated by the external excitation.*

Thus it is that the sensitive excitations awaken the activity proper to the elements of the cortical substance; that these are brought into play; and that the sensitive process, like a force which is incessantly transformed, loses by degrees its primordial character as it advances and enters a new territory.

We see then how gradually the processes of sensibility are transformed by incorporating themselves more and more with the organism; how, starting as simple physical elements, they end by becoming, in the last term of their long course, a living excitation, more and more animalized and intellectualized by the special activities of the different media which they have successively called into action. In this respect they are quite comparable to those physical phenomena by virtue of which we see the luminous rays which pass through our optical instruments become subject to the modifying influence of the media they traverse—become concentrated, refracted, unequally diffused in secondary elements, to present themselves finally to our visual sensibility, perfected, purified, separated, and with their maximum of effect.

Genesis of the Notion of Personality and of Moral

* Account of the experiments of Schiff, p. 77.

Sensibility.—The processes of sensibility have not for their sole object the transformation of external excitations; they contribute in a much more effectual manner to operations of great delicacy, which are designed to co-operate in the genesis of the notion of our *individual personality.*

It is, indeed, through the awakening of the activity of the sensibility diffused throughout the different regions of the organism—vegetative as well as excito-motor sensibility—that this notion is engendered, developed, and maintained constantly active and alive in us.

It follows, indeed, as a natural consequence of what we have already indicated, that everything in us which is sensitive—every fibre which vibrates, every sensorial plexus which becomes erethised—produces a vibration which is concentrated in the plexuses of the cortical substance, and finds in their essential structure a vast common reservoir, the veritable *sensorium commune* into which all the excitations collected in the periphery separately flow, and in which they remain latent.

The result, as regards the secondary reactions of this *sensorium,* of the general concentration in these plexuses of all the diffuse sensibilities of the organism, is naturally that all the sensibilities of the peripheral regions of the nervous system, drained from the essential structure of our tissues, of our flesh, mucous membranes, viscera—in a word, of our whole organism—and conducted along the converging nervous filaments, as the electric fluid is along the wires which transport it to a distance, inevitably travel towards the central regions of the system, towards the *sensorium commune,* where they are simultaneously distributed; and that these conceptive regions of the *sensorium* represent, as it were, at the other pole, the sensitive foci correlated to the peripheral regions in agitation.

All these modes of sensibility, whatever be their origin, are, then, physiologically transported into the *sensorium*, and there find a symmetrical region which vibrates in unison with their peripheral tonality; so that from fibre to fibre, from sensitive element to sensitive element, our whole organism is sensitive, our whole sentient personality, in fact, is conducted, transported just as it exists, as a series of isolated currents, into the plexuses of the *sensorium commune*.*

There we are represented in detail, there all our sensitive elements are condensed, fused, and anastomosed into an inextricable unity—a unity which is itself only an expression of the organic connection of the underlying nervous plexuses. There, in a word, the synthesis of all our dispersed sensibilities, which are united in a limited space and yet faithfully reproduced, takes place. There our personality lives and feels.

Here, by means of the conductility of sensitive excitations and the intervention of the nervous system, which represents in the truest sense an organ of per-

* The conductility and dispersion of sensibility in the *sensorium*, by means of the nerve-fibres, is so real, that in persons who have suffered amputation, when any irritation attacks the stump and engages the sensitive nerves, it immediately awakes and develops in the *sensorium* the old impressions in a posthumous form. It is not simply the painful state of the sensitive nerves that the patient feels, it is also the resurrection in the *sensorium* of a portion of himself, in consequence of the persistence of the conductors which formerly supported it and in which this sensitive portion of his personality was incarnate. (*See* Muller, " Physiologie," vol. i. p. 598; Sensations experienced by persons after amputation.)

fectionment implanted in the organism—something takes place quite like what we see in a camera obscura, when we see images of the external world projected upon a screen which presents only a limited plane surface. The magnifying lenses of the object-glass, the interposed media, have conducted and directed the luminous rays in such a manner that they are presented to the eyes spread out over a limited surface, with all their gradations of tint and colour, and with the relations they have in nature.

The external world is thus projected to a distance and conducted into another region than that whence it is derived, just as the sensitive excitations of the organism are drained off, condensed, and transported to a distance by the nervous apparatuses which project them into the *sensorium*, and thus permit of their being grouped according to their natural affinities.

Moral Sensibility.—When the peripheral impressions are dispersed in the plexuses of the *sensorium*, and the cerebral cell is called into play, a new series of phenomena is developed. This depends on the spontaneous reactions of the elements of the *sensorium* which are in agitation, and which vibrate in unison, and become erethised in consequence of the arrival of an external impression.

At this moment a phenomenon, quite similar to that which occurred in the peripheral regions, takes place when the sensorial plexuses are unexpectedly agitated.

This process, which leads to the transformation of the incident sensorial impression into a physical excitation, is not accomplished coldly and passively. The thousands of cerebral cells of the *sensorium commune* that have been unexpectedly awakened acquiesce in it in their own manner. They react in a specific manner, and, like their partner-cells situated at the antipodes in the sensorial plexuses, they react according to the manner in which their natural affinities have been excited. According as the excitation has gratified or wounded their profound sympathies they are agreeably or disagreeably impressed.

A new phase, therefore, at this moment appears in the evolution of sensibility, a new element comes into play, which speaks, and is excited. This is the specific sensibility of the elements of the *sensorium*, the emotivity which is disengaged from the cortical substance; and it comes into play in a necessary, involuntary, automatic manner, by the simple awakening of the elementary properties of the regions engaged. We all know how passively we receive the excitations which agitate us, and how little free we are to feel or not to feel impressions from without.

This form of sensibility which runs riot in spite of us, these plexuses of the *sensorium commune* which comprehend in themselves all the diffuse sensibilities of the organism, represent, then, a sphere of nervous activity in erethism, always living, always feeling, in the bosom of which our total personality lives and vibrates. There, in this mysterious dwelling, it is in perpetual intercourse with the perpetual movement of the operations of cerebral life. There, according to the nature of the excitations that agitate it, it finds its keenest pleasures and deepest pains—the passionate enthusiasm which exalts it, the anguish which depresses it. There vibrate the sensitive chords of our human nature, which, alternately tightened and relaxed, express the different pitches of the emotions that set them vibrating.*

The phenomena of moral sensibility, conceived, as we have just done, as a purely physiological synthesis of all the nervous activities, consist then in a series of regular processes, executed by the organism at its own expense, and resulting from the harmonic *consensus* of all its parts.

They present these two very significant peculiarities, which give them a supreme importance in the sum of the acts of cerebral life:

1. On the one hand, they are sustained and vitalized by means of former excitations; they live upon accumulated memories incessantly reviving.

2. On the other hand, they are stirred up and kept awake by the intervention of the intellectual regions, with which they are in perpetual intercourse.

1. Thus while the peripheral plexuses are only gifted with a limited power of retaining the external vibrations which have called them into action; the cerebral elements, on the contrary, have this power in a very high degree. As we shall see further on, they can store up the impressions which affect them, as phosphorescent bodies or collodion plates store up solar rays, and retain, for a greater or less period, a record of the elements which have more or less strongly affected them.

* The pages which Guislain has devoted to this subject will be read with interest; and it will be seen that he had a presentiment of the physiological evolution of the phenomena we are now describing. (Guislain, vol. ii. "Leçons sur les Phrénopathies," p. 12.)

The result of the awaking of this new property, as regards the evolution of moral sensibility, is that, once an excitation has been produced in the *sensorium*, once it has been incarnated with a special coefficient of pleasure or pain, it remains there like a phosphorescent gleam, and survives itself as a posthumous record.

Suppose an object or a person has induced in us a movement of expansion or joy, the memory of the joy perceived will survive in our *sensorium*, and will be re-awakened by the memory of the object or person who has provoked this pleasant emotion. In the same way, conversely, the memory of an insult, an injustice, a moral pain, persists in us, and remains attached to the person or the object that has been the cause of it. The emotion is united to this memory to such an extent that it is enough to think of it to cause an unpleasant emotion in us.

We know that when we voluntarily recall the image of touching scenes at which we have been present, their reminiscences evoke in us emotions similar to those we experienced at the period when they actually occurred. We know also how profoundly the anniversaries of private griefs or public calamities affect our natural sensibility.

We may say, then, that moral sensibility is engendered by the fact of the arrival and persistence of impressions in the *sensorium*. It is a phenomenon of memory, the memory of the heart, as has been said, which lives and develops itself in us, and is only sustained by means of old emotions, which, always more or less lively, are always alive and always ready to cause a sympathetic thrill throughout the sensitive plexuses of our inmost personality. Moral sensibility, then, becomes the resultant of all our joys and sorrows, and the sympathetic link which unites our present to our former emotions.

2. Moral sensibility finds also in the intervention of intellectual activity a new power, which excites it, makes it active, and maintains it in a perpetual state of erethism.

It is, in fact, most interesting to observe the important part that the intellect plays in the evolution and maintenance of the freshness of our natural sensibility.

If our sensibility finds an individual existence in the plexuses of the *sensorium*, we may say that it is enlightened, directed, educated only by the direct participation of the intellect and its manifestations. Without the intervention of the intelligence, our sensibility, with all its riches, would be nothing but an inert brute force, diffuse and completely undisciplined.

It is, in fact, the forces of the intellect, always active in the form of discernment, which make us reflect upon the choice of things or persons which have more or less affected our *sensorium*. It is because we have an acquired experience of certain persons or things that we can give them our confidence. To choose our relationships and friendships, and thus to make repeated rapid diagnoses of men and things, is an entirely intellectual operation, which illumines with the light of our reason the too often involuntary impulses of our natural sensibility.

Again, it is by virtue of the same intimate participation of the intelligence in the acts of our sensitive life, that a written page, a word, a sound, an appearance, can suddenly thrill all the emotional regions of our being in an agreeable or disagreeable manner.

When I receive a telegram or a sudden piece of news which throws me into trouble and consternation—when the reading of a comic author develops hilarity in me, it is still the direct intervention of the intellect, and the intellect alone, which excites and develops these sad or joyful impulses of my moral sensibility. It is because I comprehend—because my intellect works and immediately interprets the value of the written characters—that I remember that each word expresses a thought, and determines a sentiment of a certain pitch. It is, then, always the intellect, active and present, that in the presence of a sudden excitation, *ab externo*, awakes, and causes emotions appropriate to the external phenomena which they symbolize, to arise in the *sensorium*.*

In the same series of facts, when, for instance, in a foreign country I salute with emotion the appearance of the national flag, which is displayed before me as a symbol of my distant country, I surely do not see in it merely a piece of many-coloured bunting. No—at that moment a series of associated memories is awakened in me. I involuntarily think of a long past of glory, honour, and devotion, which is unrolled with its folds, and of the patriotic solidarity which unites me with those who defend it ; and thus, from idea to idea, from memory to memory, all the elements of my moral sen-

* This connection between the intellectual and emotional regions is so intimate, that in dreams, when the intellectual regions, abandoned to their free automatic activity, engender the strangest conceptions, we are sometimes seized with impressions of sudden terror, and overwhelmed in consequence of certain terrific apparitions.

sibility, awakened by a single physical impression of an external symbol, are thrilled, one after another, because this external symbol has awakened, in the regions of intelligence, old ideas, and national memories.

Thus, then, the activity of the intellectual regions excites, and incessantly keeps our moral sensibility permanently awake within us; while at every instant of the day, in this incessant working of all our mental activities, the intellect, present everywhere, watches over all, co-ordinates memories, regulates and stimulates the impulses of sensibility, and thus becomes the natural bridle which restrains them, as far as this can be done, within the limits of right and reason.* This is so true, indeed the energy of moral sensibility is so closely connected with the energy of the intellectual faculties, that when the latter are attacked, moral sensibility inevitably falls into decay. We often observe, indeed, that in demented old men whose intellectual faculties are considerably impaired, the impulses of moral sensibility simultaneously decay, or are more or less profoundly injured.

* It is this direct participation of intellectual activity in the phenomena of sensibility proper, that produces the different modes of feeling in men according to their different degrees of intellectual culture, their mode of life, and the hereditary conditions of organization they may present. The cultivated man will be moved by spectacles different from those which please uncultivated and gross men. The refined in intellect have their special delicacies of sentiment and modes of enjoyment which are unknown to the vulgar.

CHAPTER IV.

PERTURBATIONS OF SENSIBILITY.

Physical Pain.—The phenomena of sensibility, like all phenomena of vital activity, are susceptible of alternate lowering and exaltation, and of presenting *maxima* and *minima* of oscillation, in the interval between which their average periods are comprised.

Thus, when sensibility is locally annihilated, when the histological tissues are affected with a species of local torpor, anæsthetic phenomena present themselves. When, on the other hand, the contrary phenomena occur, when histological vitality rises several degrees to a state of cellular excitement, and the nervous elements reach a condition of continuous erethism—then manifestations of hyperæsthesia or pain occur. In these two cases phenomena connected with the natural sensibility of the nervous elements are always present, and, as it were, rise from zero to one hundred degrees.

The processes of anæsthesia and pain appear to develop like those of normal sensibility, independently of any nervous plexus which underlies them, from the simple fact of the existence of a cell capable of living and feeling.

It is certain, indeed, that in sensitive cells sensibility becomes obtuse and grows feeble under the influence of certain special conditions: chloroform makes their reaction impossible. Certain narcotic substances also appear to have a stupefying action on the sensibility of certain plants. It is certain again, that the sensibility of vegetables is perverted when they are thwarted as regards their natural evolution, and do not find in the soil with which they are furnished conditions favourable to their physiological nutrition. It is certain that they *suffer* also, as it is popularly said, and that their sensitive tissues, which are impressionable by external agencies, have to contend against wounds or with enemies of all kinds belonging to the animal and vegetable kingdoms, which under the form of parasites, oïdium, phylloxera, etc., fasten upon them and attack them even in their roots, in the very sources of life, thus inflicting upon them the same calamities we may see raging among individuals of the animal kingdom.

Pain, from the very fact that it expresses a purely vital action inherent in every living cell, vegetable as well as animal, is therefore the physiological equivalent of the individual sensibility of that same cell in conflict with the surrounding medium which impresses it painfully. It exists wherever there is a cell capable of living and feeling, and independent of the existence of any nervous element. Between the simple histological irritability of any anatomical element whatever, which is the rudimentary form under which it presents itself at first, and the most exquisite expressions of sensibility in superior beings, there are merely infinite degrees of sensitive vibrations which mark its different modes.

Just as we see a metal rod placed in a blazing furnace grow hot by degrees, and in proportion as the undulations of the caloric become more and more frequent, pass in succession through the shades of bright red, dark red, and white heat, and develop as it grows hot both heat and light; so the living sensitive cells, in presence of the excitations which affect them, undergo progressive exaltation as regards their natural sensibility, arrive at a period of erethism, and with a certain number of vibrations *disengage* pain, as the physiological expression of this sensibility super-heated to a white heat. This is so true—the phenomena of pain are so really an act of vital reaction, that not merely the awakening of sensibility but a certain tension of it, is its necessary condition. When the nervous plexus is torpid, anæsthetic, pain cannot be developed. Suffering is not a thing of the will—to suffer we must feel.

All physicians know what curious phenomena the skin of hysteric patients often presents in this respect. You may pinch them, prick them, apply burning substances to the surface of the body; the patients feel nothing save the simple contact of the substances applied; their sensitive plexuses, stricken with a species of torpor, are incapable of erection, becoming excited, and disengaging pain.

In producing local anæsthesia we obtain a similar condition of the sensitive plexuses, and prevent the evolution of pain. When the anæsthetic agent is applied it acts locally upon the individual sensibility of the nerves of the region. It chills them in a manner, hinders them from becoming heated, as regards the excessive production of painful vibrations, and maintains them at the low pitch of general sensibility. The

PERTURBATIONS OF SENSIBILITY. 117

anæstheticized regions, in fact, cease to disengage pain, while they are still conductors of sensitive impressions.

Pain being only the expression of the histological sensibility of the nervous elements risen to an extra-physiological pitch, we can understand how, being always identical with itself as regards its genesis, it may reveal itself in a different manner according to the different nature of the nervous plexus thrown into agitation.

Thus pain presents itself in various modes according as it affects such and such a sensorial plexus. If it be the retina which is affected, we know that when light is too intense its sensibility is developed to excess, and leads to a reverberation excessively painful for the *sensorium*. It is the same with the acoustic nerves, when violent and strident sounds produce contusions of their natural sensibility. The olfactory and gustatory plexuses have also their own forms of suffering, and everyone knows how painfully the contact of bitter and acrid substances, or that of fœtid emanations, affects the sensorial plexuses thus brought into play. Finally, when our viscera are attacked in their sensitive elements, we all know that they complain in their own fashion to the *sensorium*, that they reveal their suffering in a peculiar manner, and that the manifestations of pain vary with the tissues engaged, the regions invaded; that, in a word, the semeiology of pain, as regards its different characters and modes, has a special physiognomy which all physicians can appreciate.

If we now pass to the examination of the processes of pain in the central regions of the nervous system, we shall see that they are developed in a manner similar to that we have just explained, and that the morbid reactions of the *sensorium* have a method similar to that of the morbid processes of the peripheral regions.

The plexuses of the *sensorium*, in the substance of which sensitive impressions are diffused, are normally insensible, like our nerves, which, when in activity, silently transmit and elaborate sensorial impressions, without our having a notion of all their minute operations.

It is not always so. Just as the peripheral plexuses are susceptible of exaltation in presence of too-energetic vibratory excitations, or by the occurrence of a local disturbance of their habitual state of existence—so the plexuses of the *sensorium* are susceptible of excessive heating,* and of exaltation when a too-vivid peripheral impression, or a too-prolonged excitation comes to reverberate through their meshes, and thus cause them to rise to the vibratory pitch of pain.

We know that the absence of repose for the brain, prolonged vigils, uninterrupted intellectual work, moral emotions, engender a local heating of the cerebral substance, cephalalgia, and aching of the brain. The calling into activity of the cerebral cell, in an extra-physiological manner, at the same time abnormally develops its histological sensibility, and induces, as a necessary consequence, prolonged erethism and pain, in the manner we have just pointed out.

We all know by experience, how painfully a piece of taskwork which does not provoke an intellectual appetite, is done—it is an effort which the brain makes against the grain; and how, on the contrary, when the

* *See* p. 77, "Experiments of Schiff."

task is a pleasant one, there is a fascination in setting to work, and a rapidity in the execution. The natural spontaneity of the brain thus supplies the place of effort.

All those who have suffered from headache know how exquisite is the sensibility of all regions of the *sensorium*; how painful a thrill is produced by the least noise from without, the slightest shock of the thoughts which traverse the brain. They know also that silence, and sleep —that is to say the cessation of every source of cerebral excitement—are the only efficacious means for charming away these painful crises through which the sensibility of the *sensorium* has to pass.*

More than this, a comparative examination of the manner in which the central and peripheral regions of the nervous system behave in presence of anæsthetic agents, shows us a new connection between the modes in which sensibility is developed in these two opposite regions.

Thus, when hyperæsthesia appears in the *sensorium*, when the pain reveals itself either as the effect of too intense peripheral excitement (a wound, or any injury of the surface of the body), or as the effect of a persistent irritation (moral emotion, prolonged intellectual labour, etc.), we may artificially cause the level of painful over-excitement to fall several degrees, just as if we had to deal with a peripheral plexus in a condition of

---

* Just as we have seen before, with regard to the peripheral regions, that pain was only the expression of the sensibility of living tissues in exercise; so for the central regions pain is only possible in proportion to their soundness. The slow destruction of the *sensorium* by chronic disorganizations, progressively leads to the cessation of certain forms of cephalalgia. Thus we find paralytics who at the beginning of their disease have had very severe headaches, end by no longer suffering from any painful symptom.

painful erethism, and may thus to a certain extent dull the painful vibrations. It is thus that anæsthetics and stupefying drugs act when introduced as inhalations.

In operations on patients under chloroform, this agent spreading through the plexuses of the *sensorium* freezes its nervous elements, which it steeps in the same anæsthesia in which the sensitive plexuses of the skin of a hysterical patient remain.*

Painful excitations are in vain launched from the peripheral regions in the form of keen incisive thrills, when the tissues are cut; they meet in the *sensorium* only zones of cells physically modified, stricken with anæsthesia, and incapable of erection, of feeling, or of being raised to the pitch of pain.

To complete the resemblance, just as we see analgesic patients whose skin is pinched, and into whose tissues needles are thrust with impunity, witness with indifference and without painful reaction what takes place in their bodies; so we meet with a certain number of operation patients who, being capable of analyzing their sensations at the moment of operation, tell us that during the period of anæsthesia into which they were plunged, they have felt the cold of the knife penetrating into their flesh—that they have felt the

* To understand the mere mechanism of anæsthesia, it should be known that chloroform does not act simply upon the nervous elements. If we place a muscle in the vapours of ether and chloroform, or inject into a limb a weak solution of chloroform or ether, we induce rigidity of the muscle; and when we examine the change produced, we perceive that the contents of the muscular fibre are no longer transparent, but have undergone coagulation. It is to be supposed that something analogous takes place in the nerve cell, but this is much more delicate, much more sensitive to the action of chloroform, it being first to undergo coagulation. As the chloroform is carried off by the blood, the cell recovers from its anæsthesia and returns to its normal condition, as the muscle ecovers from its rigidity.

keen instrument cutting through their tissues, but that to their surprise they perceived that they did not suffer, and that the usual pain was not naturally disengaged as they would have expected. One of them told me that he experienced a surprise similar to that of a person who should plunge his hand into a burning brazier, and should naturally be astonished at not feeling the burn.

Moral Pain.—Moral pain is only the expression of the moral sensibility carried to its maximum of intensity, as physical pain is but the most exquisite form of the physical sensibility thrown into agitation. The conditions of evolution are the same in both cases, except that moral pain presents itself to us under special aspects of amplitude and intensity, which give it an expression of a persistence quite characteristic.

Thus in studying the etiological conditions of moral sensibility, we have seen how this sensibility was but a long synthesis and the resultant of a combination of the sensibility of the *sensorium* thrown into agitation with the involuntary revival of memories, and the incessant participation of intellectual activity, which always underlies its manifestations.

External excitations, as we have already remarked, once deposited in the *sensorium*, do not become extinct all at once. They survive, and like phosphorescent gleams, leave persistent traces of their passage in the nervous plexuses. On the other hand, the excitations of intellectual activity are also concerned in the process. They are always alert, always active, and by virtue of their automatic energies they reveal themselves in the shape of ideas associated with contemporary reminiscences and connected reflections; so that they also constitute, as it were, so many foci of activity capable of incessantly intensifying the movement in the plexuses of the *sensorium*.

The result, as regards the genealogy of moral pain, of the double participation of these two physiological factors—the persistence of impressions, and the incessant participation of the intellect in the phenomena of sensibility, is. this, that when the plexuses of our *sensorium* have been thus thrilled vividly to their depths, the impression so produced does not immediately die away. It becomes persistent—lives upon memories, and vibrates like the dolorous echo of a former agitation of our sensibility, to be effaced only as this sensibility becomes dulled in the region where it was primarily engendered. The shock once produced, it becomes incarnate, and perpetuates itself in us by producing the phenomena of moral grief. We cannot avoid feeling it, and suffering—each in his own manner it is true, each in a different degree, according to the delicacy or richness of the nervous elements which constitute his *sensorium*. It is no more possible to escape from a painful emotion which comes to inflict a sort of contusion upon our natural sensibility, than to escape an ecchymosis when a heavy body crushes our integuments.

On the other hand, this participation of the intellect in all that concerns us, and all that moves us, naturally becomes a species of incessant morbid excitement of our moral erethism, and perpetuates the griefs of the sensitive regions of our being. The physiological excitations which stir up and vivify our moral sensibility are, then, also those which vivify and perpetuate our moral pains.

It is because man can judge of the loss which he undergoes in consequence of the sudden ruin of his affections and dearest hopes; because he can estimate the happy memories which are fleeting; the bygone joys, the sorrows of the future, and the griefs of the present; because he finds before his mind's eye a multitude of elements furnished by his intellectual activity working automatically—that he suffers morally in his sentient being, and that the wounds of his heart, incessantly revived by a crowd of memories which assail it like so many morbid stimulations automatically arising, remain always open; that pain lives within him and preys upon him perpetually.

Vulnus alit venis et cœco carpitur igni.

Thus it is that when trouble attacks him he passes through that series of dolorous stages which lead him to slow despair, to that phase of profound despondency so often the road to mental maladies.

The moral life of an individual, his stock of natural sensibility and emotivity, is therefore kept in a condition of freshness and integrity only by the incessant activity of his memory, and intelligence, and the conscious perception of the things of the external world.

When the memory and intelligence begin to fail, and the energy of the mind to grow weak, the decadence of the moral sensibility follows that of the intelligence step by step. In a man intellectually degraded we can only count upon a low morality. And this is so true, that a person whose intellectual powers have been already impaired, either by the occurrence of diffuse cerebral congestions or by alcoholic excesses which have impaired the very substance of his *sensorium*, no longer feels moral pain according to the regular processes by which it is developed in his fellow men. The student of mental maladies frequently meets with individuals apparently reasoning with inflexible logic, and preserving a certain energy of the intellectual faculties, yet no longer having any exact notion of what is passing around them, or capable of comprehending, like every one else, the emotions of moral sensibility. If we try to convey to them a family trouble, or the loss of one formerly loved; if we seek to set some chord of emotion vibrating within them, nothing moves them. They remain impassive, and this defect of moral reaction indicates at once their dullness of comprehension, and the silence of the intellectual activity which has not normally interpreted the sense of the words and their range of significance. In this defect of sensitive reaction, we have a criterion which indicates to the observer the secret dilapidations which have occurred in the sphere of mental activity.

To sum up, it is in this special mode of evolution of the moral sensibility, in its dependence upon both ancient memories and intellectual activity, that we must look for the secret of the strong action of moral influences upon the development of diseases of the brain.

It is because man is sensitive that he suffers, and because he is, as an individual, sensitive in a certain manner, and in certain favourite directions—because he is more or less interested in the acts of his life, and conscious of what passes around him, that he suffers morally. The moral wound which is established in him, once produced, does not heal up all at once, it extends its influence, festers like a serpiginous ulcer, and being incessantly irritated by automatic impressions radiating from the sphere of the intellect, perpetuates itself, always poignant, in the *sensorium*, reviving in a thousand forms on the smallest provocation. It thus becomes, by reason of the special conditions of the medium into which it has eaten, a cause of ruin, of progressive wearing out of the mental energies, unless a profound diversion be immediately created, or a salutary method of treatment intervene to arrest disorders which tend to become incurable.

CHAPTER V.

DEVELOPMENT OF SENSIBILITY.

SENSIBILITY in living beings awakens with life. As regards histological sensibility proper, it is inherent in the primordial phenomena of the evolution of the embryonic cells; it is a hereditary legacy which accumulates incessantly, by the addition of new elements, and new tissues, in proportion as the organism completes and perfects itself.

It is by virtue of the individual sensibility of the embryonic cells that these borrow from the surrounding medium, the fluid atmosphere which bathes them, the elements suitable for their special nutrition, and that the nervous system itself appears as an apparatus of centralization and organic perfectionment.

In the first phases of fœtal life it is very difficult to fix definitely at what epoch sensibility manifests itself as a motor force; nevertheless, from the fourth month we can observe that the nervous system begins to react and to 'reveal the vitality of the different apparatuses of which it is made up.

We know, indeed, that from this period the fœtus is sensitive to the action of cold, and that we can develop its spontaneous movements by applying a cold hand to the abdomen of the mother. We know also that it executes

DEVELOPMENT OF SENSIBILITY. 127

spontaneous movements to withdraw from pressure that constrains it and brings its sensibility into play.

We may then legitimately conclude that here we have the first gleams of awakening sensibility, which from this period is transmitted through its natural channels by the nervous system, and already regulated in the manner in which it will subsequently manifest itself throughout the organism.

At birth it is the entire cutaneous sensibility, suddenly awakened by the irruption of the young being into a cold atmosphere, which determines its first startled cries, and its first inspirations. It is, then, in the sensitive peripheral regions that the first sparks which are to develop the play of the organic machinery, and those excitations of the vital knot which once set in motion will only cease at the end of life, have their origin.

From this time forth the child takes the breast of the nurse automatically, and by virtue of hereditary vital forces which already exist in a latent state in his nervous system. His organic appetites are gratified by the milk he sucks, and he feeds himself organically, like an organic cell, which borrows from the surrounding medium the materials which suit it. But at the same time he expresses the satisfaction he feels in his own manner; he smiles on seeing the breast which yields him his nourishment and life, and from that time his natural sensibility is thrown into agitation, his *sensorium* is affected. He rejoices because he remembers, because he has retained a memory of the satisfaction of his physical appetites.

Here, in these first phases of the manifestations of human sensibility, is the rudimentary formula according to which the moral sensibility of the human being shall henceforth be evolved in the course of his life, and already such as we have found it in the adult—that is to say, reducible to a purely sensitive phenomenon multiplied by the intervention of memory and intelligence.

From these first moments onwards sensibility develops rapidly.

The different sensorial foci by the aid of which it comes to life, light up, multiply, and successively attain to perfection. The child successively learns to see, hear, feel, smell, and taste. He remembers satisfactions received. He recognizes the persons who immediately surround him and load him with caresses. It was the sight of the bosom of his nurse which in the first instance excited his first smiles, and as his field of vision extends, it is the entire person of his nurse to which these same smiles appeal; then, as it extends still further, he recognizes those whom he frequently sees, and who present a pleasant physiognomy to him.

Soon, by the progressive unfolding of all the latent activities of the organic elements which come into existence, the general life of the child develops in ample luxuriance.

Moral sensibility undergoes the same developmental movement; intelligence and memory enrich these first manifestations every instant.

Henceforward the first links of family affection bind themselves round his heart, and thus become the origin of his first sentiments and emotions. He loves those who approach him, for the sake of the good things they have already done for him. He can recognize those who wish him well or ill, or who are simply indifferent to him; and thus it is that to every one who comes in contact with him, and excites his sensibility in one way or another, he devotes an appropriate memory and a gratitude proportioned to the good or evil influence he has received. He loves his parents, in the first place, because they contribute more or less to his well-being and his pleasures, and because he is in the habit of seeing them every day ; and this incessant renewal of physical impressions keeps the sentiment of gratitude in a condition of permanence and freshness in his *sensorium*. Those who are always present before his eyes are similarly present in his heart.

At another period of human existence, the most violent of the sentiments which are calculated to set all the sensitive chords of the living being vibrating—love—develops itself merely by virtue of the same physiological laws.

It is at its outset, as in the young child, the satisfaction of physical sensibility which forms the necessary prelude to it, its first stage and indispensable condition.

It is because he has been thrilled in all the elements of his physical sensibility that the living creature, at the period of love, is inevitably hurried forward, by invincible hereditary impulses, towards the being destined to be his complement and to become the physiological receptacle of his deepest joys.

It is because he has been charmed at once, in all the sensitive elements of his being, by the sight of the plastic beauties of the object of his desires, by the seductions of her speech, her voluptuous contact, and all her intellectual and moral wealth, that he is captivated and subdued. It is because all his physical sensibilities have been simultaneously awakened, and that a period of generalized erethism is developed in his *sensorium*, that he loves the object who has been for him the origin of all his happiness—that he attaches himself to her, becomes her slave, and surrenders himself altogether; just as, when he was a child, he loved, according to the measure of affection of which he was capable, the nurse who gratified his first sensuous appetites.

Thus it is that love, the concrete expression of all the sensibilities thrown into agitation, develops itself in the living being as a recognition of physical pleasures satisfied, and as a hope of their repetition; and that this sentiment, so simple in rudimentary organisms, in which sensibility is little developed, becomes complicated in the animal series in proportion as the sum of the sensitive elements multiplies, and the phenomena of moral sensibility come more into play.

In fact, in proportion as we pursue the study of this sentiment through the series of living creatures, we see that, by slow gradations, it undergoes a progressive transformation, and that in proportion as the moral influences of civilization become paramount, the purely animal physical love of savage peoples loses its primitive character, to become clothed in new forms, appropriate to the new medium in which it is developed.

Thus it is that polygamy, which is the social expression of the satisfaction of all physical pleasures, insensibly gives place to monogamy, the most perfect expression of the union of the man and woman, and a more serious guarantee for the maintenance of the family. This regular form of love, which is an epitome of the most delicate perfections of human sensibility, concentrates upon a single head the sorrows and joys of the past and the hopes of the future, and thus cements the permanent ties consecrated by the customs of common life. It inevitably engenders, in every degree of the social scale, spite of the numerous shortcomings by which it is dishonoured, those natural acts of devotion and self-abnegation for the common work of progeniture, and that whole series of respectable sentiments of which the domestic morality of monogamous peoples offers most striking examples.

As a man advances in life, his sensibility becomes gradually lessened—the senses become dull, the sight loses its sharpness, the skin its impressionability by external agents.* A sort of general slackening of all his functions impends over the living creature thus arrived at the retrogade phases of his evolution.

This condition of diminution of the peripheral sensibility is reflected in a similar manner upon the sensibility of the central regions. Moral impressionability and emotivity lose their energy as a man grows old. He is less and less interested in external things capable of exciting his mental activity. He is less sensitive, less impressionable, less curious as to knowledge and feeling, and at the same time his intellectual faculties are simultaneously impaired. Memories of the past, like enfeebled phosphoric gleams, persist for a certain time, to the exclusion of more recent remembrances, but, in

* In old persons the skin atrophies very remarkably, and in a great number the skin of the *derma* is so attenuated, that by pinching up a fold in the dorsal region of the hands I have often been able to observe that it has become so thin and translucent that the circulation in the subcutaneous capillary plexuses might be seen, as in the foot of a frog.

the end, even they too are extinguished, so that, the circle of bygone things narrowing by degrees, the individual feeds his *sensorium* only with the current operations of life. Material life with all its necessities—eating, drinking, and sleeping, becomes, little by little, the favourite occupation of organisms in the period of decadence; and as to moral sensibility, the old man, an egotist with few exceptions, is reduced to vegetative life, and becomes once more a child, caring no longer for those who care for him day after day. He forgets his old friends, and the most natural family affections, for the sake of the newest comer, and succumbing more and more to the interested demands of his personality, he arrives, as regards moral sensibility, at a true anæsthesia which reflects the languishing condition of the elements of his nervous activity.

BOOK II.

ORGANIC PHOSPHORESCENCE OF THE NERVOUS ELEMENTS.

CHAPTER I.

INTRODUCTORY.

I HAVE proposed to apply the term *phosphorescence* to that curious property the nervous elements possess, of remaining for a longer or shorter time in the state of vibration into which they have been thrown by the arrival of external excitations—as we see phosphorescent substances illuminated by solar rays continue to shine after the source of light which has illuminated them has disappeared.

We know, indeed, now, thanks to the works of modern physicists, that the vibrations of the ether, in the form of luminous undulations, are capable of being prolonged by phosphorescent bodies for a longer or shorter time, and thus surviving the cause which has produced them.

Nièpce de Saint-Victor, in his researches on the dynamic properties of light, has arrived at results much more precise and unexpected; since, in a series of reports,* he has shown that luminous vibrations may be to some extent garnered up in a sheet of paper, and remain as silent vibrations for a longer or shorter period, ready to appear at the call of a revealing substance. Thus, having kept in darkness some prints previously exposed to the solar rays, he, several months after this insulation, succeeded in demonstrating, by means of special reagents, persistent traces of the photographic action of the sun upon their surface.

On the other hand, the daily practice of photographic reproduction by means of dry collodion, is an irrefragable demonstration of the aptitude which certain substances gifted with special elective sensibility have for preserving persistent traces of the luminous vibrations that have for a certain time affected them. In fact, when we expose a plate of dry collodion to the luminous rays, and several weeks after such exposure develop the latent image it contains, we produce a resurrection of the persistent vibrations and obtain a record of the absent sun; and this is so true, in this case of persistence of a vibratory movement which has but a limited duration within which it must be seized, that if we pass the prescribed limits and wait too long, the movement gradually becomes enfeebled, like a source of heat which cools and ceases to be able to reveal its existence.

This curious property, which inorganic substances possess, of preserving for a longer or shorter period a species of prolongation of the impressions which have first set them in motion, is found once more under new

* "Comptes-rendus de l'Académie des Sciences," Nov. 16th 1857, vol. xlv. p. 811, and March 1st, 1858, vol. xlvi. p. 448.

forms, with special phenomena, it is true, but essentially the same, when we come to study the dynamic phenomena of the life of the nervous elements.

These also are gifted with a sort of organic phosphorescence, and are capable of vibrating and storing up external impressions, of remaining for a certain time in a sort of transient catalepsy, in the vibratory state into which they have been incidentally thrown, and of causing the first impressions to revive after the lapse of time.

We all, indeed, know that the cells of the retina continue in a state of vibration after an excitation has ceased. It has been calculated by Platau that this persistence of impressions may be estimated at from thirty-two to thirty-five seconds.* To this persistence of vibrations, and that special retentive force which the nervous elements possess, is due the fact that two successive and rapid impressions become confounded, and thus give a continuous impression: that a live coal whirled round at the end of a string gives the impression of a circle of fire: that a disc, painted with the colours of the spectrum, when in rotation gives only the sensation of white light, because all its colours are confounded and form for us an unique resultant, which is the idea of white. All those who occupy themselves with histology know that after prolonged work the images seen in the focus of the microscope live in the *fundus* of the eye, and that sometimes, after several

* The duration of impressions upon the retina is much longer than that of the action of light. According to Platau, the duration of the consecutive impression increases in the direct ratio of that of the primary impression (?in the direct ratio of its *intensity*). Thus the consecutive image of a strongly illuminated body may be kept in the eye for a very long time. (Muller, " Physiologie," vol. ii. p. 355.)

hours' work, shutting one's eyes is sufficient to cause them to reappear with great distinctness.

It is the same with auditory impressions. The auditory nerves preserve for a long time the trace of impressions which have set them vibrating. After a railway journey, we hear, for several hours after arrival, the noise of the rattling of the carriage. A musical air, and certain favourite refrains, involuntarily resound in one's ears, and that often in a most disagreeable manner.

After long musical *séances*, says Dr. Moos (of Heidelberg), the sounds persisted for fifteen days in one patient, and in another, a professor of music, for several hours after each lesson.*

The gustatory plexuses also seem capable of thus preserving the trace of agreeable or disagreeable impressions which have affected them, and the intensity of the impression is sometimes lively enough to produce, retrospectively, either a secretion of saliva when the mouth waters at the thought of something nice, or, in other circumstances, a sensation of nausea when the substance has produced an unpleasant sensation.

The impressions of general sensibility, olfactory sensibility, etc., appear to present analogous phenomena.

This species of histological catalepsy, which to some extent polarises the nerve-cells in the situations in which they have been immediately placed at the time of their first impression, is not merely a unique phenomenon, which is met with in the peripheral regions of the nervous system; it is also met with still more fully developed in the central regions of the system, where it appears with such pronounced and fixed characters

* "Annales Médico-psychol.," vol. ii. p. 121, 1869.

that we might say that it governs the manifestations of automatic life in the spinal cord, and directs those of psycho-intellectual activity in the brain.

In the different segments of the spinal cord the persistence of impressions reveals itself very evidently in the accomplishment of all those co-ordinated movements which, not being a part of the hereditary patrimony of the motor apparatuses of the organism, are therefore acquired by habit, being the direct product of education.

We know that the greater number of the rhythmic movements we execute in most bodily exercises—dancing, fencing, playing on musical instruments—are methodical movements which we never accomplish (except the first time) by the intervention of the will; that they are the effect of long apprenticeship; that they are only acquired by exercise, the force of habit, and the imitative tendency we have, to reproduce patterns presented to us. Now, our muscles can move in such marvellous union according to given indications —our movements can be harmoniously combined in accordance with the operations to be accomplished, only by virtue of the latent aptitude of the excito-motor cells of the spinal cord for preserving records of the impressions that have first thrown them into agitation—for remaining for a longer or shorter time in the primordial condition first imposed upon them.

It is, then, our first impressions that vibrate in us like distant echoes of the past, and serve as a stimulus to the excitations of automatic life. It is they that, always alive, always faithful to themselves, are incessantly disengaged in the form of *unconscious reminiscences*, regularly rhythmic motor manifestations, which faithfully reproduce the impression of the primordial excitation.

It is the same persistent excitations, condensed in the sphere of automatic activity, that in certain morbid cases, when the regions of the *sensorium* and conscious perception are temporarily closed to impressions from without, excite those very curious harmonic movements accomplished by certain somnambulists, which take place *motu proprio*, by the simple calling into activity of the automatic regions which act of themselves, and exhibit externally a series of *unconscious reminiscences*. In connection with this subject, Mcsnet has lately reported a most interesting case—that of a soldier, who, having received a shot in the head, afterwards suffered from very strange symptoms.

This man was subject to a species of somnambulistic crises, in consequence of which his *sensorium* was to a great extent cut off from all external impressions. He ceased, more or less suddenly, to enter into contact with the surrounding medium, and then, while in this condition, would walk about, go and come, and if anyone endeavoured to direct his movements in any definite manner, the impulse was inevitably developed in the direction of former excitations preserved in the state of unconscious reminiscences in the plexuses of his automatic activity.

Thus, for instance, on putting his walking-stick into his hand, the touch of it reminded him of his gun, and he would then place himself in a position as though he were present at a battle. If a pen were put into his hand, the precise movements necessary for tracing written characters were unconsciously produced in him.

These motor excitations were automatically developed in the store of latent reminiscences grouped according to a primordial arrangement, and producing, as it were, phosphorescent gleams of the past; as we see in decapitated animals similar movements excited through the preservation of the automatic activity of the spinal cord.*

Legrand du Saulle has reported a case which is somewhat analogous to the preceding. It is that of a young somnambulist, a ropemaker by trade, who, if seized with a fit of somnambulism when twisting his rope, would continue the operation he had begun, even while asleep.†

In my own wards I had a patient, still young, who had been for a long time attached to the Salpêtrière, as an assistant in the linen-room, being employed to fold the clothes and roll bandages. In the last years of her life this woman, completely blind and paraplegic, presented the following phenomena. While lying on her back, if any one put into her hands an unrolled bandage, or even the end of a cord, the touch immediately awoke in her reminiscences of her former work, and she began automatically to make a rolling motion with her hands, without knowing what she was doing, as though she had been a piece of machinery.

We may then assert that the nervous plexuses of the spinal cord preserve in their minute structure (like the peripheral nervous plexuses, the retina among others) records of the impressions which have previously excited

* Mesnet, "Sur l'automatisme de la mémoire et des souvenirs." (Union Médicale, 1874, number 87.)

† "Annales Médico-psychol.," 1863, tome I. p. 89. (Legrand du Saulle, "Le somnambulisme naturel.")

them, and that these persistent records thus become like a series of fixed autogenic excitations, designed to act at a long range, to radiate to a distance, and thus to produce a series of reactions quite similar to those to which they at first gave rise. These phenomena of motor reaction, which take place merely through the calling into play of the organs of automatic life, are capable of spontaneous evolution, and of producing a repetition of certain habitual movements without any participation on the part of the conscious personality, which is absent for the moment.*

In entering upon the study of the cerebral activity proper, we shall see what an important part this property which the nervous elements possess of retaining a record of former impressions, plays in the operations of the life of the brain, and in what varied forms this organic phosphorescence, always identical with itself, always present and distributed throughout the nervous elements which compose the tissue of the brain, performs its functions.

It is diffused throughout all the agglomerations of cells, which are like so many active foci of phosphorescence, but unites into a single resultant which concentrates all the sparse activities of the cerebral cells. It thus becomes, under the denomination of the general faculty of memory, a true synthesis of one of the primordial properties of the nervous elements.

The elements of the cerebral substance, the uncon-

* See an account of experiments made on the body of a decapitated animal, in connexion with the development of manifestations of automatic life, in a direction determined by previous habit, and of the persistence of certain movements directed to a certain end. (Ch. Robin, "Journal de Physiologie," Paris, 1869. p. 90.)

scious agents of the manifestations of our psycho-intellectual life, work in silence at the operations which they accomplish in common. They associate together, with their manifold properties, in one harmonious effort, corresponding with one another by the mysterious channels of their anastomoses, and without our knowledge preserve in their minute organism posthumous prolongations of past impressions. They act simultaneously to produce the phenomena of memory, and separately give off reminiscences, as illuminated bodies give off the luminous waves they have stored up in their substance; this marvellous power of the cerebral cells, which depends on the favourable conditions in the midst of which they live, being maintained in a condition of perpetual vigour so long as the physical conditions of its material constitution are observed, and so long as it is associated with the vital phenomena of the organism.

The phenomena of memory, thus looked at as a necessary consequence of a fundamental property of the nervous elements, enter directly into the mechanism of the different regular processes of cerebral activity. They may consequently be looked upon from the successive points of view of their genesis, their evolution, their mechanism, the diverse phases they pass through during the life of the individual, and the functional disturbances from which they are liable to suffer.

CHAPTER II.

GENESIS AND EVOLUTION OF MEMORY.

IN order that the processes of cerebral activity which constitute memory shall be evolved according to their natural laws, it is necessary that the peripheral regions of the system which collect and transport sensorial impressions, on the one hand, and the central regions which transform and absorb them, on the other hand, shall be reciprocally in suitable conditions of physiological conductility and receptivity.

1. It is indeed in the peripheral regions, in the midst of the ultimate nervous expansions, that the activity of the central regions finds its regular food. Thence it is that all the stimulations destined to set them in motion proceed.

When an external excitation is reverberated to any point whatever of their essential structure—whether it be a sonorous wave thrilling through the acoustic expansions, or a luminous wave becoming extinguished in the regions of the retina, or any direct stimulus which sets in vibration the·sensitive nerves of the skin and mucous membranes —immediately this purely physical excitation is transformed on the spot by the peculiar action of the nervous plexus in erethism. It absorbs it, transforms it into nervous vibrations, and to some extent animalizes it by incorporating it with the organism.

Now, since the peripheral nerve-cells, as we have said, retain in themselves, like phosphorescent gleams, the record of those stimulations which have first set them vibrating, the result is that these persistent impressions become, without our knowledge, like a store of latent peripheral reminiscences, which hold the partner cells of the central regions in a sort of persistent vibratory sympathy. They in their turn assist the action of the central memory, and thus become a means of physiological reinforcement designed to vivify and maintain its activity.

This solidarity between the peripheral and central regions of the system is so real, that when the former fail, the functionment of the central regions is at the same time interrupted.

When the sensitive peripheral regions are in a state of anæsthesia central perception ceases. There is no persistent reminiscence in the *sensorium*, because the trace of the persistent peripheral impression has not been registered. Touch, pinch, excite the skin of a hysterical patient in any way you please, if the eyes be closed, she will retain no remembrance whatever of the cutaneous excitations, because her peripheral nervous plexuses being stupefied, will not have been able to transmit to the *sensorium* anything that has taken place in their internal structure. I have often seen general paralytics, attacked with transient anæsthesia of the gustatory and pharyngeal nerves, bitterly complain to me that they had not been given a particular dish at their meal, I having been present when they had partaken of the food which they declared they had not received. Then again, the absence of sensibility in the peripheral region causes the sensorial impression not to be absorbed on the spot, nor directly transmitted to the central regions by its habitual channels.

In order that the sensorial impression shall produce the desired effects in the plexuses of the *sensorium*, and shall be clearly perceived, it is necessary then that the peripheral plexuses, which are its true gates of admission into the organism, shall be in a condition of receptivity and peculiar erethism, that their natural sensibility shall be directly awakened, and that there shall be on their part an active and prolonged participation when the stimulation from without arrives.

Every one knows, indeed, that a slight and fugitive impression leaves but insignificant traces of its passage; that an incessant repetition of the same impressions is necessary, in order that they shall be retained in a stable manner ; and that it is only by dint of forgetting, that we come to have certain details present in our minds which escape us and which it has been necessary to learn again and again. The repetition of the same peripheral impressions, the repeated view of the same objects, the hearing of the same sounds, become therefore indispensable fundamental conditions of the preservation of reminiscences ; and from this point of view the reminiscences emanating from the sensorial *plexuses*, *the memory of the senses*, as they are pedantically called, are the most energetic stimulations of mental memory.*

On the other hand, in order that the impression per-

---

* All those who have pratically studied anatomy know how necessary it is frequently to review certain regions of the human body to know them well ; and that it is only after having seen, touched, and dissected, that we succeed in fixing in our memories the different details we have studied.

GENESIS AND EVOLUTION OF MEMORY. 145

sistent in the peripheral plexuses shall produce a durable impression in the central regions, the preceding conditions of centrifugal impression are not the only ones necessary. It is necessary that there shall be something more on the part of these same central plexuses of the *sensorium*—an effective participation or intimate association of their sensibility with the peripheral excitations which thus throw it into agitation.

At the moment, indeed, when the external impression sets the peripheral sensorial cells vibrating, these are affected, according to the different modes of their natural sensibility. They are sensitized in a different manner, according as the excitation is agreeable or disagreeable to them. In the first case a sensation of pleasure accompanies the external impressions, in the second case a sensation of discomfort; so that the nervous element coming into play with its latent activity, transports to the *sensorium*, not only the announcement of the arrival of the external excitation, but at the same time the special notion of pleasure or pain related to each excitation.

Every former impression, every reminiscence that slumbers within us, remains there from the moment it has been perceived, stored up with a specific coefficient which recalls to us the joy, the pain—or even the indifference of these same peripheral plexuses at the moment when it was incorporated with them and when it began to live in their own life.

We all know that the reminiscence of physical pain, and corporeal chastisement, so lively in animals that are in training, is for man one of the surest guides of his conduct, and a most faithful warning to

L

avoid faults which will inevitably provoke their recurrence.

We know, conversely, that reminiscences of agreeable impressions, and those which have given us most joy, are also those which have the deepest roots in us, and that in fact different states of emotivity, associated with the arrival in the *sensorium* of such and such a group of external impressions, are what perpetuate themselves with the greatest tenacity. They thus become, as regards the desires they excite or the aversions they beget, the natural pivots around which all human activities gravitate.

2. We have just seen the mode of genesis and transmission of persistent sensorial impressions, at the moment when they are begotten in the peripheral regions of the system—let us now see how they are received in the plexuses of the *sensorium*, and what reactions they provoke as their consequences.

The connections between the peripheral plexuses and those of the *sensorium* are so intimate that, so soon as an impression has been produced in the former, their partner central regions immediately enter into unison with them. There is a nervous condition of similar pitch which harmonizes one part with another, and whenever the primordial impression has been sufficiently intense, and sufficiently prolonged, whenever there has been an effective participation of the nervous plexuses laid under contribution, the partner plexuses of the *sensorium* sympathetically associate in their excitations and enter upon a concordant period of erethism. The incident excitation arrives then in the plexuses of the cortical substance, purified and animalized by the peculiar metabolic action of the nervous plexuses in the womb of which it is incarnated, and then, transforming itself into a psychic excitation, it develops the latent energies proper to the cerebral cells, imprints itself upon them, and perpetuates itself in them in the form of persistent vibrations, like a phosphoric gleam of the external world.

Thus it is, that this mysterious property which the nervous elements possess—that of persisting in the vibratory condition in which they have been placed—is here again found consistent with itself throughout the different stages traversed by the sensorial excitations; from the peripheral regions where it reveals itself in so indubitable a manner (as in the persistence of impressions on the retina), to the central regions, where it acquires characters entirely dependent upon the multitude of elements which serve to maintain it.

Thus it is then, that external impressions of all kinds, the diverse emotions we have felt, become finally attenuated in the plexuses of the *sensorium*, and in the form of persistent vibratory thrills become the posthumous expressions of impressions and past emotions which remain alive in us when the primordial excitations have long ago disappeared.

Sensorial excitations, when they are diffused in the plexuses of the *sensorium* and fix themselves there in a persistent manner, do not usually remain there in the state of vague, uncertain impressions. They go further, penetrate more deeply into the recesses of cerebral life, and when they are sufficiently lively and often enough repeated, they penetrate even into those inmost regions where the notion of *conscious personality* is elaborated, and thus become conscious reminiscences of ancient emotions that have thrilled us.

Thus it is that, as regards the phenomena of memory, our inner personality is seized upon by the same process by which it was seized upon on the arrival of sensorial impressions; only that these impressions which call it into activity prolong their action, implant themselves in the organism, and become, as it were, a vibratory echo of the past. It is thus then, that the reminiscence of anterior excitations perpetuates itself in the *sensorium* with the particular coefficients of joy or sorrow that have presided over their genesis in the peripheral regions, and thus a series of emotions related to each of them becomes developed, and perpetuates itself in the central sensitive regions of our organism.

The phenomena of psychical and moral activity, understood as we have previously explained, perpetuate themselves in a similar manner, and develop incessantly, by the mere calling into activity of the two fundamental processes of the nerve-cells—sensibility, and that peculiar retentive power, organic phosphorescence, by means of which they prolong the vibratory excitations which have first set them in motion.

In the domain of intellectual activity it is still the same force that underlies most of the dynamic operations to which this activity gives birth.

It is, indeed, because he remembers, because his sensibility has been impressed in a special manner, and this impression is persistent in him, that the young child, from the first instant of his life, expresses his inner sentiments. It is because he remembers, that he recognizes external objects and names them with an

GENESIS AND EVOLUTION OF MEMORY. 149

appropriate word, which he has retained in his memory from having heard it. It is by means of the persistence of acoustic impressions, preserved in the state of sonorous reminiscences, that he speaks, and that his phonetic expressions are applied to each surrounding object.

It is also by the same means that he learns to trace written characters, which he recognizes as the symbolic expression of absent objects, and that he reads aloud, transforming each written character into sonorous concordant expressions which he knows to be their equivalents.

There are always at the bottom of these different operations of the intelligence, persistent sensorial impressions which direct the processes in evolution, and vibrate like a faithful echo of the first impression. It is the same with that admirable faculty which the human being possesses, the power of translating into verbal expression his emotions and the thoughts which pass through his mind. It is because man has learned that each word expresses an external object, a thought, a sentiment, and because this acquired notion, preserved by daily use, is maintained in him in a state of permanent freshness, that he speaks, addresses his kind, and is understood by them. It is memory—the accumulated reminiscences always present to the mind—that forms the basis of his language, and thus becomes the inexhaustible store in which he finds the means of expressing what he feels and what he thinks.

CHAPTER III.

THE MEMORY IN EXERCISE.

BESIDES those phenomena of memory into which the human personality more or less enters, there exist a whole series of similar acts which represent processes of memory to some extent incompletely developed.

These are those phenomena in which sensorial excitations, not having carried their action as far as the plexuses of conscious personality, remain in the condition of sterile materials, not perceived by the *sensorium*. Like those dark ultra-violet rays of the spectrum, which though not perceptible to our eyes, have nevertheless a real existence, they remain silently accumulated in the plexuses of the cerebral cortex, and only await the presence of an exciting cause capable of causing them to start from their obscurity.

Thus, we all know that during the period of our diurnal activity, there are a host of various impressions which assail us on all sides, and even strike redoubled blows upon our sensitive plexuses, yet to which we pay no attention. The multifarious noises of carriages rolling around us all day, finally come to be unperceived by us and indifferent to us. We know also that when we give ourselves up to an absorbing intellectual work, the ticking of the clock beside us strikes

THE MEMORY IN EXERCISE. 151

in vain upon our ears, and yet our acoustic nerves have been again and again set vibrating without our having a notion of it.

Onimus has made a very curious observation in connection with this class of ideas. A man who was walking began automatically humming an air, being very much surprised by its having come into his head. It was only accidentally that he perceived that the air had been suggested to him by a wandering musician who was playing it on his instrument as he passed by, and whom he had not perceived.* This man in humming the air echoed an auditory impression, an unconscious reminiscence.

We all know that in examining a picture, or landscape, or a histological preparation, we first passively see the whole, and that certain details when we are not prepared for them at first escape us; and if a person, after we have gone to a distance from the object we have examined, retrospectively calls to our notice certain peculiarities of the object, we are quite astonished that we have remarked them, and that we recognize in ourselves the existence of certain impressions which have remained silent.

It is by means of unconscious impressions which persist in the brain that the activity of our spirit, in the automatic work which takes place in the act of reflection and meditation, is maintained.

It is thus that the unexplored sides of certain questions in suspense are made clear by the juxta-position of old impressions which have arisen. A sort of automatic appeal is made to revived impressions which have

* Onimus, "Journal d'Anat. et Physiol.," de Robin, p. 551, 1873.

some connection, and which come, as new factors, to enlighten our judgments with a number of new ideas.

The symptomatic study of mental maladies presents, as regards the subject, phenomena which are often very curious. We sometimes meet persons who have received an excellent education—ladies, young girls, living in the best society, above all taint of impurity, who, when seized with an attack of cerebral excitement, utter the grossest words, quite strange to their ordinary vocabulary.

Evidently, in these cases, the phenomena can only be explained thus:—That in walking in the streets or in public places, these gross phrases have unconsciously impressed them, and have remained in the state of latent memories buried in the cerebral tissue; and that it is because of the morbid over-activity of the regions in which they are stored up that they are discovered and leap to light.

Local Memories.—It results from the anatomical arrangements, to which we have so many times directed attention, that the different groups of sensorial impressions have each a special territory of distribution in the different regions of the *sensorium*, and that consequently there are in the human brain inequalities very clearly distinguished as regards the part devoted to each particular order of sensorial impressions. (Figs. 5 and 6.)

It follows then from this inequality of development of similar regions in different individuals, that there exist special aptitudes for the reception of the different kinds of sensorial impressions. Thus it is that one person, whose optic cerebral regions are abundantly provided with well-endowed lively nerve-cells, will be fitted for clear perception of the external world—surrounding objects, with their colours and relations; that another (Fig. 6.—14, 15), whose acoustic cerebral regions are largely developed will be predisposed to appreciate all the shades and delicacies of musical harmony; while a third will have such and such an aptitude according to the preponderance of such and such a region of his brain; and that thus, the special sensorial impressions, finding within such or such a circumscribed locality conditions of soil more favourable, agglomerations of cells more dense and more lively—these impressions will leave more enduring records, more vivid remembrances, and from this very fact richer stores of materials for fertilizing the psycho-intellectual activity in such or such a direction.

We are thus led to the conclusion that there are in the phenomena of memory, taken as a whole, certain peculiarities, by virtue of which this memory is more or less vivid in such or such an individual as regards such or such a cerebral operation, and that thus there are a certain number of local memories very clearly determined, each having, in a manner, an autonomy as independent as the generating sensorial impressions with which it is intimately associated.

Association of Memories.—The study of the brain has shown us that there are isolated regions designed to receive and elaborate independently isolated groups of sensorial impressions. The study of the cortex, on the other hand, has shown us that if there be a certain functional autonomy, as regards the dispersion of impressions, this autonomy is neither complete nor definitive, seeing that examination of the nervous tissue of the cortex, proves that this tissue forms a continuous whole throughout all its extent, a unity as complete as that of the cutaneous surface—so that the excitations perceived at a given moment, in a certain region of the *sensorium*, are nevertheless liable to be disseminated at a distance, and to associate the different regions of the cerebral tissue in their vibration. (Figs. 1 and 6.)

Thus, for instance, when, in the presence of a picture, a landscape, or any object which may catch our eye, the special regions of our brain which elaborate optic impressions are thrown into activity, it is only homogeneous optic impressions, and nothing but optic impressions, which are active in a determinate region of the cortex.

When, on the other hand, my view extends over a landscape, or over a flower-bed balmy with fragrant emanations ; when I am present at a theatrical representation in which the splendours of the *mise en scène* equal the magnificence of the musical harmony, my brain is no longer excited by a homogeneous stimulus ; it is assailed by a series of simultaneous impressions which come in a crowd and impress themselves all at once upon the *sensorium*.

These simultaneous impressions—optic, olfactive, acoustic, received at the same moment, and in several circumscribed localities at the same time, constitute a series of contemporaneous souvenirs which are created and implanted in me ; and, henceforth, those vibrations which were born together, and were simultaneously conceived, will represent, in the general series of my reminiscences, a definite group, of which the elements, united by the bonds of a mysterious federation, will all live with the same life, anastomose one with another, and recall one another as soon as one link of the chain is struck.

Thus it is that the sight of even a corner of the landscape I first saw, or of the flower-bed that gratified my sense of smell, will recall to me the odour of the plants that I had pleasure in smelling, and even the emotions that I experienced at that very moment; and inversely, these perfumes accidentally inhaled at a later period, will evoke in me automatically a reminiscence of the place, the flower-bed, where they were simultaneously inhaled. Thus it is, again, that the sight of such or such a theatrical decoration will remind me of the piece of music heard in its presence, and that, in the same way, if under other circumstances I hear the strains that have impressed me, I shall feel awakened within me reminiscences connected with it, which will represent to me the decorations and the ocular spectacle in the presence of which I heard the musical sounds for the first time.

By taking more and more complex examples, we find that in the ordinary phenomena of cerebral activity, not merely are binary, ternary, or quaternary groups of sensorial impressions juxtaposed and imprinted upon the *sensorium*, but many multiple agglomerations are created within us and proceed from all the sensibilities of the organism successively and simultaneously laid under contribution.

Thus the pleasures of gastronomy may easily be allied, as Brillat-Savarin has so well explained, with all other pleasures; the seductions of physical pleasure, which are the synthesis of all the sensibilities of the organism in agitation, leave in the *sensorium* traces all the deeper, and memories all the more vivid, because they represent a series of juxtaposed, successive, partial impressions, which multiply each other, and mutually co-operate, so that they appeal to one another, associate in a thousand forms, and, thus implanted by their innumerable roots in the *sensorium*, become like a series of conjugated foci for exciting in it a condition of erethism.

This curious property which sensorial impressions received at the same time possess, and which constitutes as it were natural families among them, is a great resource in the education of the intellect, and the methodic cultivation of its faculties. When a series of memories, a series of ideas, of experimental facts and scientific principles, has been imprinted on it, it admits of their being artificially evoked, contenting itself with an appeal to the first in the series of memories, which is in a manner at the head of the line.

Thanks to this connection between our particular reminiscences, the intelligence incessantly acquires new riches, and may at a given moment, by means of its automatic activity, seize upon these riches and make use of them.

Thus, when from observation of a clinical fact, for instance, we have learned that, a case of acute rheumatic arthritis being given, this condition of effusion into the joints is accompanied by a similar manifestation as regards the heart, these two impressions henceforward form in the mind two united memories ; so that, the first being given, the second immediately arises, and *vice versâ*. In the presence of a patient with rheumatism we think of a cardiac affection, and conversely in the presence of an old affection of the heart, we interrogate the patient as to his rheumatismal antecedents.

When I have learnt from the experience of my masters that lesions of the posterior roots and posterior columns of the cord are accompanied by defective co-ordination of movements, ocular disturbances, sharp sudden pains in the limbs, gastric troubles, etc., I have anastomosed in my mind by study a series of memories associated one with another and forming a sort of federation; so that when one happens to be isolatedly evoked—when, for instance, I see a patient with special ocular troubles, I spontaneously think of defective co-ordination of his movements, the existence of sharp sudden pains, etc.

What here takes place in an order of facts clearly determined, with regard to a series of phenomena methodically regulated, is constantly and regularly renewed in us during the period of our daily activity.

When every region of our brain is in erethism, we all know how memories appeal one to another; always following the same series in their method of appearing, without our being able to command them. It is sufficient to see an object or a person—to hear a name pronounced accidentally, to smell an odour—in order to feel arising within us a series of ideas which arose at the moment when this impression was at first perceived by us. We all know how frequently in current conversation a word—a simple sound—causes the primitive direction of our ideas to diverge. Many persons indeed thus lose sight of the point of departure, accidentally led away by a passing reminiscence which introduces divergent thoughts, and insensibly causes them to turn aside from the subject with which they have started.

Do we not all know that when we wish at a given moment to evoke a particular reminiscence, and fear that the distractions of our current life will cause us to forget it, we mentally attach the object to some sign, which thus becomes for us the clue that recalls it to our mind. Every one has his own mnemonic on such occasions, and we all know that it is sometimes a knot made on a pocket-handkerchief—an object which must necessarily pass through our hands, a ribbon fixed upon our garments, a visible mark designed to catch our eye mechanically, to which we have recourse, in order to cause the reminiscence we wish to evoke to leap forth in the natural course of events.*

* In the practice of mnemonic methods we know that our end is the association of a series of reminiscences difficult of retention, by the help of strange combinations of words; these words, easy to retain by reason of their strangeness, containing in themselves analytic solutions of the difficult points we seek to fix in the memory.

CHAPTER IV.

DEVELOPMENT OF THE PHENOMENA OF MEMORY.

THE general faculty of memory, the organic phosphorescence of the nervous elements, is liable to present great modifications, according as it is considered at the different periods of the development of the human being. It goes through successive phases, which are merely more or less direct reflexes of the histological properties of the cells, by means of which it reveals itself.

In young children the cerebral cells are endowed with special histological characters; they are flabby, greyish, flexible in a manner; they are, moreover, from the dynamic point of view, virgin to any anterior impression. The sensorial excitation that affects them at that age must therefore imprint itself upon them more readily, since it finds them in a state of vacuity, their power of retention not being as yet put to the test.

On the other hand, in the first years of life the cerebral substance is in perpetual exercise and organic development. New elements are perpetually being added to the old ones, and as the new are most probably derived from their predecessors, we are led to conclude that the daughter-cells which appear, borrow from the mother-cells which give them birth an inevitable bond of relationship, a species of hereditary transmission of the different states of the mother-cells from whence they spring. It is, then, probable that the primordial cells, which give birth to all the generations of daughter-cells that appear in the course of cerebral development, transmit to their descendants the special sensitive properties, the specific degrees of phosphorescence, with which they were animated at the moment of their origin; and that it is in these intimate connections between cell and cell, in these mysterious bonds of relationship, that we must look for the secret of the perennial character of certain memories. Thus it is that certain impressions received in our early childhood become the common patrimony of certain families of cells, which maintain them in a state of freshness, incessantly vivifying them by a sort of permanent co-operation.

In the young child the impressionability of the cerebral substance is such that it retains, *motu proprio*, all the impressions that assail it, as passively as a sentized photographic plate that we expose to the light retains all the images that are reflected on its surface.

Visual and sensitive impressions are the first to be inscribed upon the *sensorium*.

The child sees objects and persons that interest him, within a restricted circle. These first impressions captivate him, and he keeps the remembrance of them, individually recognizing each person or thing. Little by little, auditory impressions coming into play, he hears sounds, which are vague at first, without comprehending or interpreting them; and insensibly, by the effect of the activity of the brain, from the persistence of the impressions and the notion of them that he acquires, he comes to recognize that these determinate sounds answer to precise objects, which are always the same, and which in some way interest his personality.

Little by little, this work of cerebral culture being pursued without cessation, new acquisitions are incessantly registered in the *sensorium*. The different modes of sensibility awakened bring with them new ideas and new remembrances, and at the same time excite appropriate reactions. The regions of intellectual activity begin to make more and more use of the excitations which come from the surrounding world to erethise them.

At this happy age the child retains what he sees, hears, tastes, without trouble. The strangest words, complete phrases that he does not comprehend, abstract proper substantives, pieces of poetry, the operations of mental calculation, leave in him persistent impressions which are perpetuated and registered in a stable manner. It is this special period of complete absorption, which we might call *the age of substantives*, that represents in the history of the development of the human being the first rudiments of intellectual activity, as, in the history of the development of humanity in general, the stone age represents the first outlines of human labour.*

* Substantives play a principal part in the evolution of thought and speech. They are the primordial *data* around which the verb and the other parts of speech group themselves. They are the elements that underlie the combinations of human thought. The facility with which they disappear from the memory in certain cases of cerebral disorganization suggests the thought that from the first periods of intellectual development, they are really received and stored up in isolated territories of nerve-cells which serve them as a substratum, in the form of persistent sensorial impressions.

In the adult the elements of cerebral activity in a condition of complete development are endowed with all the energies they are capable of assuming. They do not now behave as they did in the young child during the period of his evolution, as far as regards the preservation and storing up of external excitations.

The period of saturation begins for the cerebral cell. The power of retention of external excitations is already on the brink of decay. New acquisitions of heterogeneous elements which do not form a portion of the circle of youthful knowledge become very difficult, if not impossible. We know how painful the labour of learning a foreign language, so easy for the young child, becomes for the adult; how rebellious the memory is as regards the registering of new words; and with what an expenditure of intellectual force we retain the vocabulary of languages with which we were not familiar in childhood. We also know how blunt, even in the domain of common things, the retentive power of our memory, and consequently our powers of application in general, become, if we have to learn things that are quite new to us; and how, for instance, we with reason look upon it as impracticable to acquire a special technical education, and commence a new career after forty years of age.

At this period of life first impressions still faithfully persist in the memory, but nevertheless they have a tendency to diminish in intensity, and it is necessary to vivify them by incessant labour, to stimulate them anew by placing the cerebral regions where they are stored up in identical conditions, by similar impressions of equal intensity, so as to prevent their becoming extinct; just as we keep up a fire by continually supplying it with fresh material.

As the entire human frame begins to suffer from the effects of senescence, which occurs in different individuals at very different periods, the cerebral cells, like the other elements of the organism, suffer a premature decay.

They grow old histologically; they become more or less infiltrated with fatty granular matter; they cease to be transparent, shrivel up, and from a dynamic point of view insensibly lose a portion of their sensibility and their special retentive power; so that, as foci of organic phosphorescence, it may be said that they are extinguished within certain circumscribed localities of the cerebral cortex, and consequently cease to preserve a record of their first impressions. Thus it is that the general phenomena of mental activity undergo a perceptible decay proportional to the sum of the cerebral elements superannuated. In the aged, memories sometimes disappear in an isolated manner; sometimes those which are not maintained by regular exercise become extinct; sometimes the general faculty of memory fails altogether, and in its decay involves the progressive blunting of the most lively sentiments.

A strange phenomenon now occurs—we perceive, contrary to what *à priori* would seem most probable, that in old persons, as in patients with dementia, old memories remain the freshest and most vivid, while recent facts, impressions which occur at the very moment, are unperceived and treated as if they did not exist. It is probable that at this period of life, the cells of the *sensorium*, altered in their essential constitution, have become lazy, and incapable of erecting themselves in the presence of recent external impressions; and that this state of torpidity of the elements of the *sensorium* for new excitations, leaves the field free to the older ones which, not being obscured by more lively impressions, continue to vibrate without opposition, and thus perpetuate the last phosphorescent gleams of a far-off past which is dying.*

* Thus, in some old persons in dementia, from the mere fact of the non-absorption of recent impressions into the *sensorium*, the notion of the passage of time is completely annihilated.

From the fact that the daily work of the absorption of new impressions has ceased, the individual remains fixed in one spot, as it were, in a cataleptic state, with the ideas and preoccupations that he had at a given moment of his existence. Thus, we see a great number of patients who, having been some ten or twelve years in an asylum, still keep the ideas they had at the moment of their entrance, without having an idea of the passage of time; and who, if asked how long they have been there, will speak of two or three years.

CHAPTER V.

FUNCTIONAL DISTURBANCES OF THE PHENOMENA OF MEMORY.

THE manifestations of memory, looked at as we have just done, do not then present themselves merely as a collection of simple phenomena, nor as the direct resultant of the impression made upon the plexuses of the cortical substance by an external excitation. They consist in true physiological processes, which have an origin and a regular evolution throughout the nervous system. They demand the active participation of the cerebral cell; and to be regularly executed they must obey certain organic necessities, and the inevitable conditions of integrity and co-operation of the organs through which they effect their complete development.

When, therefore, any disturbance whatever occurs either in the essential vitality or in the constitution of the organic elements which they lay under contribution, the processes of memory are *ipso facto* disjointed, and that faculty is thus maimed in one or other of the operations that constitute it.

Thus there are circumstances in which that property which the nervous elements possess, of retaining a record of external excitations which have formerly impressed them, attains a condition of extreme and permanent exaltation. This vibratory phase of their existence perpetuates itself and becomes a species of unsubduable erethism.

All phyiologists, indeed, have recognized the important part a sudden emotion, such as terror or the sight of an epileptic attack,* plays in the production of convulsive seizures; and I have further pointed out that violent impressions may remain stereotyped in certain individuals attacked with general paralysis, and that the shock caused in the *sensorium* may be very vivid, since it is capable of manifesting itself for several consecutive months in a species of cataleptic condition, imprinted upon the countenance, and upon the attitudes of the body.†

The symptoms presented by the automaton whose interesting case has been reported by Mesnet, come under this class of facts. There are in such cases persistent impressions, which have been formerly accumulated in the *automatic sensorium*, which continues to direct the excito-motor processes without participation of the conscious personality.

Van Swieten, who was seized with vomiting on coming upon the dead body of a dog which exhaled an insupportable stench, chanced upon the same spot some years afterwards. The memory of what he had experienced produced the same disgust and the same consequences.‡

This class of morbid phenomena is always developed by virtue of the same physiological processes as those

* See Luys, "Actions réflexes cérébrales," p. 83. Morbid phenomena resulting from a persistent impression. (Paris, 1854.)
† Luys, *loc. cit.*, pp. 73 and 87.
‡ "Annales Medico-psychol." 1851, p. 242. Fact cited by Parchappe.

which regulate the manifestations of normal activity. There are latent and silent stimulations which, by reason of certain conditions which have presided over their impression upon the organism, remain more vivid than others, and which, by virtually becoming incessantly-active stimuli, produce a discharge of nervous force, either in the form of interrupted convulsive currents, in that of continuous motor currents (cataleptic condition of the muscles), or in that of sympathetic reactions from the side of vegetative life (vomiting, etc.).

In other circumstances, we have no longer to deal with an isolated phenomenon, revealing itself by definite manifestations, and reflecting as before the deviations of a normal process regularly accomplished. We observe, in fact, manifestations of quite a different kind, which reveal themselves by a species of exaltation of the psycho-intellectual regions, which preserve and store up external impressions in a very vivid manner, and when the cerebral elements have risen above their usual pitch, manifest their new condition by an unexpected super-activity quite contrary to the habits of cerebral life of the individual.

We see patients, indeed, gifted with very ordinary intelligence, who, when in this phase of cerebral erethism, will improvise, make quotations, associate new ideas with extreme rapidity, say witty things and make puns—things they are quite incapable of doing when in their ordinary vital condition.

Michéa cites the case of a young butcher whom he observed in the Bicêtre, and who, under the influence of an attack of mania, recited whole speeches from the *Phèdre* of Racine. During an interval of calm, he said he had but once heard the tragedy in question, and that, spite of all his efforts, he could not recite a single verse.

Van Swieten cites from the same author the case of a young workman, who, never having dreamed of making verses, during an attack of fever became a poet and inspired. Perfect speaks of a lunatic, who, during his delirium, expressed himself in very harmonious English verse, although previously he had never shown any disposition for poetry. Tasso is said to have worked better during an attack of mania, than in his lucid intervals.*

Finally, in other circumstances we observe phenomena of an entirely inverse character. Far from being phenomena of over-excitement of the memory, they are those of dislocation and clouding over.

Persons thus affected, more or less completely lose the faculty of retaining certain memories; either through the destruction of certain circumscribed regions in the cortical substance,† or through the progressive destruction of its elements.

Similarly there are certain persons with dementia who, being affected with partial amnesia, forget the date of the day and year in which they live; they do not know their way, lose themselves in the streets, and yet they are still able to sustain a certain amount of current conversation. Others, on rising from table forget they have had their dinner, and order it to be served up. Others, after receiving a visit from their relations or friends, and

---

* Michéa, "Annales Médico-psychol.," 1860, p. 302.

† Voisin has pointed out a case of amnesia with softening of the cerebral substance. The patient had lost the memory of objects, and had forgotten names and substantives. If a spoon were presented to him he could not tell the name of it, but showed by his gestures that it was for eating soup with. ("Sociétié Anatomique," 1867, p. 342.)

conversing with them, when the visit is fairly over—an hour afterwards—retain no definite impression about it, or else make mistakes; when, for instance, they have received a visit from their daughter, they will say they have had one from their grandfather, etc.

There are others again who, although enjoying a certain portion of their faculties and the capacity for speaking regularly, lose little by little the memory of proper names, then that of substantives, then of verbs, and make mistakes in orthography. Cuvier, in his lectures, mentions the case of a man who had lost the memory of substantives, and who could form sentences very well, with the exception of names, which he left blank.*

It is curious to remark, as J. Falret has done, that in this process of decay which takes place, the human mind in despoiling itself of its wealth, loses it chronologically in the order in which it has accumulated it. Thus it is the remembrance of proper names which is first extinguished; these, as we have previously remarked, p. 161, representing the first periods of the work of the intelligence in ascending evolution. Then come common names, adjectives and verbs, which represent a more advanced degree of the perfectionment of the faculties, when the child has begun to express his will by means of appropriate verbs.

Thus in these periods of progressive decadence the processes of memory being gradually deprived of the materials by means of which they effect their manifestations, cease to be regularly evolved; amnesia advances further and further, and we see individuals thus affected

* "Annales Médico-psychol.," 1852, p. 305.

quite incapable of registering present impressions, preserving no remembrance of what passes around them, forgetting the past, and becoming more and more incapable of expressing their sentiments and wishes, in consequence of the progressive wearing out of the organic apparatuses that serve for the evolution of the processes of memory.

BOOK III.

AUTOMATIC ACTIVITY OF THE NERVOUS ELEMENTS.

CHAPTER I.

INTRODUCTORY.

THE automatic activity of the nervous elements, like their histological sensibility, is merely one of the special forms of their peculiar vitality.

Diffused, in a similar manner, in its most simple forms, through the most elementary organisms, this automatic activity is perfected, and amplified, in proportion as it is distributed through more abundant and more dense agglomerations of cells, which are at the same time endowed with a more intense vital energy.

It reveals itself in its most simple forms, as a histologic property of the free cells, the white corpuscles of the blood; of that series of cells with mobile prolongations (vibratile cilia, spermatozoids), whose automatic energy is manifested in such characteristic amœboid movements; and finally of isolated masses of protoplasm.

As we ascend in the zoological series, we perceive that the manifestations of automatic life consist not merely in purely local phenomena, in which the histologic elements accomplish the natural phases of their evolution *motu proprio*, but in the exhibition of new dynamic properties. The histological elements, then, secrete, as it were, at the expense of their substance, peculiar autogenic excitations, and project them to a distance in the form of a continuous or interrupted current, thus acquiring a species of power of radiating to a distance the vital forces they have locally evolved.

Thus we see electric fishes accumulate, in special tissues of their organism, the electric force which they emit, for the purpose of defence, in the form of discharges regulated by a voluntary excitement.* Thus also we see the superior animals condense in the nervous plexuses of their organism stores of motor influence, to be distributed through the peripheral regions in the form of complex manifestations of voluntary motor-power, or of the motor-power of vegetative life.

The operations of automatic activity are, then, generally characterized by a series of processes inverse to those of sensibility. In fact while the phenomena of sensibility are usually characterized by centripetal currents which pass from the peripheral regions where they are conceived towards the nervous centres, the phenomena of automatic activity, on the contrary, are marked by currents with a centrifugal direction. With the former they complete the cycle, and reflect outwards the excitations which arrive from the external world through the sensitive regions.

Now if we consider the phenomena of automatic activity, from the point of view of their relations, and

---

* "De la substance électrique ou élément anatomique caractéristique du tissue électrogene." Ch. Robin, "Journal de l'Anatomie," 1865, p. 510.

AUTOMATIC ACTIVITY OF NERVOUS ELEMENTS. 173

their connections with the nervous system, we see that for them also, for the organic force which excites them, the nervous system similarly plays a perfecting part, that it amplifies them, gives them its own energy, places at their disposal its conducting filaments, and thus enables them to reach their highest point of perfection.

They follow indeed, step by step, the progressive stages of development of the nervous apparatuses with which they are connected. Thus, in the peripheral regions of the system, where the phenomena of vegetative life take place by means of automatic forces alone, the nervous elements—represented by the unicellular sympathetic ganglions, which are like so many little outposts in the web of the tissues—interfere only occasionally to regulate the different rhythms of the local circulations.

In these distant regions the automatic life of the individual elements reigns without contest. It is local activity that rules here; and a sort of complete decentralization characterizes the life of these regions.*

Little by little as we approach the centres a progress towards complete subordination takes place in the distribution of the living forces of nervous activity. Thus, if we pass from the ganglions to the medulla, we observe that sensitive phenomena are distributed in certain regions, and motor phenomena in others. Sensibility and automatic activity, which were vaguely fused together in the peripheral ganglionic masses, are

* Unicellular ganglions, or ganglions composed of a few cells, have long been observed in the intestinal coats, in the bladder, and in the walls of the vessels. (Legros, "Thèse d'agrégation sur les nerfs vaso-moteurs," Paris, 1873, p. 14.)

here distinctly separated, and exercise their functions regularly by means of nervous cell-territories specially adapted for a determined end. This is still not all;—in the brain this principle of the progressive perfectionment of physiological work by the complexity of the apparatuses by which it is accomplished, becomes more and more evident; so that automatic activity is revealed not only in the phenomena of motor-power, but also in the manifestations of psycho-intellectual activity.

Wherever, in fact, the phenomena of nervous life are developed, they appear not only with those general characters of individual sensibility and organic phosphorescence which we have hitherto recognized as being the essential attributes of every living nerve-cell, but with a new co-efficient in addition—that property, so characteristic of automatic activity, the capacity for spontaneous vibration, if their natural sensibility, previously aroused, be thrown into agitation, and for radiating and projecting to a distance the expression of that histologic sensibility thrown into agitation—at first in the form of an automatic reaction completely independent of the existence of the nervous system, and subsequently in the form of nervous discharges.

The automatic activity of every living cell is, then, nothing but the spontaneous reaction of its individual histological sensibility, evoked in some manner or another.

This special form of the vitality of the nervous elements we are now about to consider. We shall thus see that these automatic activities, together with sensibility and organic phosphorescence, become the fundamental elements of cerebral activity ; that they associate one with another in a thousand ways, and combine to produce the most complex operations of cerebral dynamics; and that they always underlie most of the operations of cerebral life.

CHAPTER II.

GENESIS AND EVOLUTION OF AUTOMATIC ACTIVITY.

Spinal Phenomena. — The phenomena of automatic nervous life reveal themselves, as we have said, in their simplest elementary form in the mysterious operations of vegetative life, while the sympathetic ganglions, scattered through the web of the tissues, and connected with the central regions by their connective threads, locally govern the phenomena of the local life of the different cell-territories, and act as little eccentric centres which hold in subjection the purely vegetative phenomena.

In the centres, in the purely spinal regions, the manifestations of automatic life again reveal themselves in an independent manner, as though they had a special autonomic character in each of the particular regions of the spinal axis.

This automatic activity is so vivacious in the minute structure of the grey plexuses of the spinal cord, that it persists of itself, exercises itself *motu proprio*, apart from all participation of the superior regions of the encephalon; and each segment of the cord, considered as an independent ganglionic centre, may also, even when distinctly isolated, function regularly and give rise to co-ordinated reactions.

In fact, if we cut the spinal cord of living animals into separate segments, as Landry has done,* we shall find that each segment will isolatedly give rise to a series of independent motor phenomena; and as long as the blood-currents continue to feed the cells, and these can store up new force after each discharge, and continue to live their morphological life as before, they will continue to produce nerve-force, and inevitably give rise to regularly co-ordinated phenomena, according to previously established habits.

Moreover, the experiments of Ch. Robin, made upon the corpse of a decapitated criminal,† have shown that the automatic activities of the spinal cord in man may in similar circumstances continue to exhibit undiminished energy and power of co-ordination, in the form of regularly associated movements with a definite object (such as movements of defence made by the hand after a cutaneous excitation), performing these with as much regularity as though the brain had directed them.

We have also true types of automatic reactions in that series of excito-motor processes which succeed each other without a break throughout the *medulla oblongata*, the region of the vital knot, and in which the cells of this region, like the indefatigable workmen of our great manufactories, work incessantly night and day for the regular maintenance of the foci of innervation of the heart and the respiratory muscles—and this without break or halt, our whole life long, without the intervention of the conscious personality, and merely through the permanence of the automatic forces.

* Landry, "Traité de paralysies," Paris, 1859, p. 48.
† Ch. Robin, "Journal de l'Anatomie," Paris, 1869, p. 90.

It is, moreover, a remarkable fact that this automatic power of the spinal organs is so great; its participation in all acts that we primarily accomplish with the concurrence of our conscious will is so effective and regular, that little by little it succeeds in gaining ground in the domain of our conscious dynamic operations, obtaining a greater and greater importance by means of prolonged exercise, and finally ruling over them more or less.

We all know that those partial movements we accomplish in tracing written characters, and in playing musical instruments, are at first executed and followed out with the participation of the conscious will, and that little by little, as exercise, as it were, oils the automatic machinery, this comes into play on the smallest excitement, like a well-constructed mechanical contrivance, automatically reproducing the movements learnt, with a neatness, co-ordination, and correctness, all the more perfect because the conscious personality plays a less distinct part in the process.

We all know, more or less, that the action of writing certain phrases, and above all that operation which is the somatic expression of the conscious personality *par excellence*—that of affixing our signature to a sheet of paper (which indicates the passage of the conscious will through the hand that expresses it) insensibly becomes an operation which escapes our attention, and which, like certain common phrases that we unconsciously pronounce, takes place of its own accord, simply through the apposition of the pen to the paper, and by reason of the coming into play of mere excito-motor activity.

We therefore see what an enormous part the phenomena of automatism are called on to play in the manifestations of nervous life, since we already know that these not only regulate the essential phenomena of vegetative life, but in addition play a most important part in calling into activity the great mechanism for the maintenance of the human machine, such as the motor-power of the heart and respiratory apparatus—in a word, the phenomena of visceral life ; and that, more than this, they enter into the processes of purely psycho-intellectual life, which have need of their intervention to project outwards their extrinsic manifestations, and escape from the mysterious regions where they have been primitively conceived.

CHAPTER III.

AUTOMATISM IN PSYCHO-INTELLECTUAL ACTIVITY.*

IF, now, we enter upon the physiological study of cerebral activity proper, we shall see in what complex forms this curious property of the nerve-cell reveals itself, and in what an infinite number of combinations it is capable of taking part.

It is principally in the perceptive regions of the *sensorium*, and those that are the seat of purely intellectual phenomena, that the manifestations of intense automatic life are most distinct.

In fact, what takes place within us when an external impression suddenly thrills us, when we find ourselves touched in the sensitive regions of our being, by the sight of an affecting scene, or a spectacle that charms us, or by the hearing of music which pleases our ears, is this: immediately, by reason of the elementary properties of the *sensorium*, which are at once called

---

* These cerebral phenomena of automatic activity have been for the first time described and very explicitly demonstrated by Baillarger, both in communications made to the Academy of Medicine, and in a series of articles in the "Annales Médico-psychologiques," under the title of "*Théorie de l'Automatisme et de l'exercice involontaire de la mémoire et de l'imagination.*"

"The more I observe lunatics," he says in this remarkable work, "the more I am convinced that it is in the involuntary exercise of the faculties, that we must seek the point of departure of all forms of delirium." ("Annales Médico-psychol.," voL vi. p. 188. *Idem*, 1856, p. 54.)

into play, sensibility is awakened, and develops itself into the sense of satisfaction, and this external impression, stored up in the vibratory condition, persists in us, and becomes a durable memory. But this is not all; these persistent impressions, transformed into durable memories, do not remain there as mere barren stores; the automatic activities of the nervous elements which have come into play are now evoked.

It is, in fact, as we have seen, sufficient that a certain series of cerebral cells shall have simultaneously undergone a series of sensorial impressions, in order that they shall form among themselves a species of mysterious association, united by the ties of contemporaneous impression. If, then, we happen to experience any excitation whatever, visual, auditory, or olfactory, the appeal of the first in the series, by virtue of these mysterious associations immediately causes the others to spring up; former memories reappear, and so blind and inevitable is the communicated movement, that this is effected without any conscious participation of the will. It does not depend upon us to incite or direct it; it follows its route by virtue of its peculiar affinities and regular anastomoses, as automatically as the sympathetic and excito-motor actions that are propagated through the plexuses of the spinal cord.

These phenomena, of the association of former memories following upon a recent impression, repeat themselves at every instant of cerebral activity. It is sufficient for us to come fortuitously upon one external object to think of another, which has either direct, or indirect and artificially maintained relations with the former.

Reading has no other rational basis. It is the memory of the thing signified, incessantly evoked by the graphic sign, that causes us to adopt automatically, with each graphic sign perceived by the understanding, ideas of which such signs are but the conventional expression.

In conversation ideas follow upon, and evoke one another in quite an automatic fashion. We think, without wishing it, of a thing outside of the subject in question, and, automatically, we are drawn away from the principal thought.

In assemblies we frequently see certain orators deviate by degrees from the subject under discussion, through the action of the automatic forces of their minds, which always lead them in the direction towards which they are biassed—that is to say, towards the regions of predilection, where their favourite thoughts have developed a species of persistent erethism. These automatic forces, which guide human thoughts in a certain direction, are so inevitable, and so apt to pass through a certain regular orbit, that, the character and oratorical habits of such and such a person being given, we may infer, *à priori*, that at a given moment he will express such and such a thought, or pronounce such and such a phrase.

In public lectures there are professors who, speaking volubly, repeat annually the same phrases, and the same words, at the same periods, and this without its being done voluntarily. More than this; it is notorious that at certain examinations, the examiners in any given subject repeat again and again the same questions; and this logic of the automatic cerebral activity is so real, that those interested have instituted a course of questions designed to calculate in advance the automatic direction that the mind of their examiner will follow, and to anticipate the questions he will put to them.

Every one knows in fine, that it is enough to set certain loquacious individuals going at a favourite subject, to make them immediately unfold all their ideas upon the theme, repeat the same things and recite the same adventures, and this in a manner as monotonous as automatic. Of this class, old soldiers, huntsmen, and travellers, are accomplished specimens, and each of us may recall similar examples in the circle of his acquaintances.

The automatic activity of the cerebral elements, when it has been too strongly over-excited, may reveal itself in certain circumstances in a more intense manner, with more vivid colours; thus assuming a special character without there being, properly speaking, delirium, since the conscious personality still looks on at its morbid condition, like an involuntary spectator.

Thus I may here cite a few fragments of a letter written by a young man, who after too prolonged work, gives a frank account of his impressions and the automatic determination of his mind to work, in spite of him.

This young man had been for several days engaged in making calculations of compound interest, which had caused a great tension of his mind. One evening, after dinner, he was about to go to sleep, when, as he says, "Without the slightest encouragement on my part, in a state between sleeping and waking, waking

I may well say, for my mind having worked beyond its powers all day, struggled obstinately against the corporeal fatigue which strongly incited me to sleep. On which side was the victory? On that of the mind. For without intending it, and having need of the greatest calm and repose to which I could attain, *I began, without the smallest volition on my part, to calculate and go over again exactly the same problems as when in my office.* The cerebral machine had been set in motion too violently to be stopped, and *this involuntary work went on in spite of me, and in spite of and against all the means I endeavoured to employ to cause its cessation*—that is to say, from about three-quarters of an hour to an hour and a quarter."

Common Sense.*—These phenomena of automatic activity are not only developed in the living being, considered as an individual, in a completely unconscious manner, but besides, by a species of diffuse generalization, they are repeated in similar individuals in an identical manner, and throughout space and time provoke in all human brains associations of ideas, and acts connected according to a general and common rule, as similar as though they emerged from a central region which gave them a single impulse.

It is, in fact, very curious to observe that there are among all human beings, modes of feeling, of judging of things, and of reacting in consequence, which are everywhere the same. Moral phenomena, in fact, occur in a manner as necessary as if we had to do with purely physical acts.

* See the complementary details of the question in the Chapter on the Judgment (pp. 291, 292).

PSYCHO-INTELLECTUAL ACTIVITY. 185

Thus, just as over all the surface of the globe, since men have existed, they move their forearm in the direction of the articular surfaces, in pronation and supination, and bend the articulations of the knee and the leg, and the head, in an unchangable manner, and in a predestined direction—so in the circle of ideas, in the gamut of sentiment, in the mode of reacting of the human *sensorium*, there are universal consonances, which throughout time and space present characters of eternal immutability.

The history of ancient literature shows us that in the same situations human beings have always felt, and always acted in an identical manner. In every page of their tragic or comic works, we find that common fund of immortal truth and judicious reflexion, which will be eternally current and applicable at every epoch. Similarly, if we consider humanity throughout space, we find that the civilized nations of the extreme East, the Chinese and Japanese, have of themselves in their long social evolution *automatically* invented the same processes of government and administration which have been for centuries contemporaneously employed in our old Europe.

Human brains, therefore, everywhere and always react in a common and identical manner in presence of the external excitations which impress their *sensorium*. Each, more or less, represents a prism of the same composition, exposed at the same angle to the same incident rays of light which traverse them. Each undergoes the action of the same rays, receives them through its substance in an identical manner, according to a common process, refracts them in a similar manner, and disperses them, after they have produced in each identical phenomena of elementary decomposition.

We thus arrive at the opinion that there is in humanity a sort of general arrangement of ideas and sentiments, by virtue of which all men automatically take the same direction in the same definite circumstances, and judge of surrounding things in an identical manner. It is this natural aptitude that we all possess for vibrating in unison with others, in presence of an external situation, for *refracting* external impressions in a fashion identical with that of our fellows, that causes us to have within us that notion of right, according to which our judgments and actions should be unconsciously directed. There is, then, a common rightline, a regular high-road which is, in a measure, the common meridian line along which the emotions, judgments, and actions of human beings are directed; and it is this inner notion, that we carry within us, which constitutes the rule of good sense and *common sense*.

The complete man regularly constituted should, then, in presence of fixed determinate emotional situations, react in an appropriate manner, make the same reflexions, experience the same attractions, and the same repulsions that his fellows experience. This is the happy point of contact which unites all humanity in the same joys and the same sorrows, associates it, under whatever latitude and at whatever epoch we consider it, in the same enthusiasms, the same sympathies and the same aversions.

Every theatre-goer has felt himself moved by the pathetic situations, and has associated his bravos and his tears with those of his neighbours. Every one of us, in solemn moments of the national life, has felt himself thrilled by the general excitement caused by those poignant patriotic emotions that the men of our generation have experienced in sorrowful alternation. Every one who stood upon the Boulevards of Paris in 1859, when the French army marched past, returning from the campaign of Italy, must have participated with all his heart in the general intoxication of victory; and every one who stood on those same Boulevards a few years after, among anxious and over-excited crowds, when all our disasters were announced, must have felt all hearts beat in unison with his own, and his secret sorrows reflected in all faces.

Communication of Automatic Activity to others.— Automatic activity works in human brains according to laws so inevitable and energies so involuntary, that we may count upon it at a given moment, consider it as a living force in the static condition, and excite it without the agency of volition, as we see, for instance, bodies electrified in a certain manner act at a distance upon neighbouring bodies, and modify the dynamic conditions of the electric forces latent in them.

The cerebral automatic activity develops itself also at a distance, passing from one individuality to another by the intervention either of speech, writing, or gestures, which excite the *sensorium* of the individual addressed; and the excitement, once communicated, is propagated from point to point, through the plexuses of the cortex in a continuous manner, by the mere automatic forces of the nervous elements, which disengage their latent energies.

Thus it is that human speech provokes in the *sen-*

sorium of any one who hears it involuntary reflexions, which traverse the brain, and finally produce a unison between him who hears and him who speaks. The art of persuasion has no other physiological *raison d'être* than the setting in vibration of the sensitive cords of the emotional regions of the *sensorium*, and the direct or indirect neutralization of previous prejudices. It is by this process that the act of causing laughter at the proper time, and of turning aside the attention by exciting unexpected sentiments is often a means of disarming one's judges.

It is by setting in motion the automatic forces latent in human brains, that great orators get possession of an attentive audience, subjugate it, and excite in it involuntary ecstasies of emotion and enthusiasm; that great writers develop a whole series of unconscious emotions through which their moving recitals hold us spell-bound; that a word or a phrase evokes a whole series of involuntary ideas, which give rise to a crowd of reflexions and emotions, corresponding to those they wish to inspire in us. It is by virtue of the same general laws of communicated emotion that the periodical publications of the press, by daily percolating through the minds of their readers, give an automatic direction to their ideas (human laziness being so fond of ready-made phrases), and produce in those who enjoy them that fixed mental direction they unconsciously acquire.

The same automatic tendencies of the human mind to provoke co-ordinated associations of ideas, thoughts and emotions, connected with other thoughts and emotions by the mysterious links of former relationship, are visible in every-day life, and, by means of words of double meaning—transparent allusions, which, in connection with one word, make us think of other words—produce the most unexpected effects, and the most unforeseen mental suggestions.

People, in fact, who in their conversation handle double meanings with art, know very well that, by underlining a word, by an inflexion of the voice, a look, a gesture, they will awake in the minds of their audience a series of ideas and emotions of a nature different from that indicated by their words. The simple phrase of allusion, when perceived in the brain in the form of a phonetic impression, follows, as it were, two parallel routes—one natural, apparent, traced by the word itself; the other roundabout, divergent, traced by the intonation and gesture. There result thus from these simultaneous processes, which are propagated through the cerebral tissue, various series of unconscious reactions, which, in the form of memories, associated ideas, and different sentiments, are successively awakened. Hence the unexpected, vivid, and piquant relations between certain ideas that provoke hilarity, and certain distant thoughts which may cause the fibres of our inner emotivity to vibrate in a more or less indirect manner.

What more simple, apparently, than to speak of a cradle to a young girl, and yet what more cutting, since one is sure to see her sensibility in agitation betray itself by the blush of modesty?

A vulgar proverb, in the same circle of ideas, says that "We must not speak of a rope in the house of one who has been hung."

The ancients, at the door of the lupanars, used to inscribe these words: "Cave canem," etc., etc.

It is in these processes, which plunge as it were by multiple roots into the fruitful stores of our memories and emotions, present and past, that dramatic literature finds its most powerful machinery. How many pathetic, and, more often still, how many comic, scenes are produced by nothing but the apparent contrast between the visible situation of the personages on the stage and the gestures and intonations of the actors, which appeal to quite another genus of ideas, thus automatically provoking, by this very fact, bursts of laughter or movements of terror—even in situations which are apparently far from inspiring gaiety or terror.

It is always the automatic activity of the cerebral elements that comes into play in those different conditions, provoked in the *sensorium* by means of plays on words and certain well-made puns.

It is, in fact, in consequence of the unexpected association of two opposite ideas that the hilarious paroxysm is produced in us.*

Reflexion of Automatic Activity.—One of the most interesting facts as regards the phenomena of automatic activity is this: that they are not only maintained

---

* In certain morbid forms of cerebral activity this automatic tendency no longer reveals itself (as regards opposite ideas suddenly associated) by similar words. It is by simple assonances which appeal one to another and group themselves together automatically. Thus a patient, described by Parchappe, with great mental volubility, often in her speech formed associations of ideas after this fashion. "On dit que la Vierge est folle, on parle de *la lier*, ce qui ne fait pas les affaires du département de *l'Allier*." On being told to make *charpie* she said that she did not know how. This was insisted on, the physician adding: "Je vous dis *d'en faire*." She answered: "Il ne fait pas bon dans *l'enfer*."

PSYCHO-INTELLECTUAL ACTIVITY. 191

by means of the incessant influx of excitations which come from the external world, impinge upon the *sensorium*, and demand its active participation, but that, in addition, they reveal themselves of their own accord, old memories forming in us, as it were, so many autogenic foci, which kindle themselves. From this it results that, by means of this prolongation of former excitations, automatic activity feeds itself, maintains itself locally, and develops itself in the form of meditation and reflexion at the expense of the stores accumulated in the past, which thus become the aliments of its incessant activity.

We all know that, when we have to come to a resolution, we have, as it is said, need for reflection, for maturing it in our mind ; that is to say, must give it up to the automatic activity of our mind, which takes possession of it, reacts in consequence, and causes new ideas, unexpected thoughts, unforeseen points of sight, which give it more weight, to arise. Night, it is said, brings counsel ; that is to say, in consequence of simple repose the cerebral elements have recovered their proper vitality, and have become more fit to develop their natural energies in presence of the resolution in question. Thus it is that the automatic forces of the brain, concentrated around a circle of definite ideas, develop themselves automatically, provoke the intervention of new elements, and finally create quite new methods of seeing and considering things. And what is well worth attention is that all this series of marvellous phenomena develops itself *motu proprio*, and outside of the conscious personality, which looks on at this subtle work, and is as powerless to excite it when it slackens, as to restrain it when it is developed in excess!

Automatism in the Sphere of the Psychic Activity Proper.—The automatic energies of the cerebral elements, as we have just seen, play a principal part in the processes of sensorial perception, as in those of intellectual activity proper. If we now pass to the examination of the phenomena of purely psychical activity—that is to say, of those which are characterized by moral sensibility and emotivity—it is not without surprise that we see that these same automatic vital forces reveal themselves here also with clearly distinguished characters, and that while always active, always identical, under the most diverse forms—either under the name of involuntary temptation, irresistible impulse, etc., they always betray the inner secrets of the emotional regions of the *sensorium* where they have originated, even in presence of the conscious will, which is powerless to regulate their manifestations.*

The labour of life is an incessant struggle between the acts of conscious volition and the automatic impulses of the emotional regions of our being. Ordinary language is rich in metaphors which express in appropriate forms what is unalterable and inevitable in this special domain of our mental activity. These phrases: *impulses, enthusiasms of the heart, sentimental biases, spontaneous outbursts of tenderness*, are trite and even silly expressions by which we have always expressed those manifestations of our sensitive nature, in which

* Emotional sensibility develops itself so involuntarily, that in the theatre, even when we know that all that is there represented is but fiction, the simple sight or hearing of pathetic scenes suffices to set our restrained sensibility in vibration, and spite of us causes our tears to flow.

there is an involuntary and unconquerable element. There are many persons to whom we feel ourselves involuntarily drawn by the captivations of their personality, and many others who, on the contrary, drive us away by a sort of repulsive radiation that they project to a distance. How often in the gamut of tender sentiments a single look has sufficed to throw the whole being into commotion, and to excite all the sensitive fibres! How often, in contrary circumstances, a menacing and imperious glance has sufficed to strike the individuals upon whom it has been darted, as with a thunderbolt, and fix them immovably to the spot! We know, indeed, that both love and hatred, from the very fact that they express different conditions of our *sensorium* in agitation, are quite automatic and unconscious sentiments. They are inspired and experienced, not commanded by the intervention of the human personality.

And it is remarkable that, just as in the sphere of intellectual phenomena there is a necessary logical order according to which they succeed one another, so there is similarly a logic of sentiments and passions which imposes itself on this series of purely moral phenomena of the natural sensibility, and which, at a given moment, follows its regular course in the heart of man, like the series of ideas which are logically connected in his mind.

It is more or less profound knowledge of these spontaneous reactions of the human sensibility in presence of such a given circumstance, that enables great writers to know point by point, and express with precision, and put into the mouths of their personages, natural expressions of the passions which are to be developed in them. It is because there is a logical order in the evolution of the sentiments and passions, that we can *à priori* infer the effects produced upon our fellows by a good and happy piece of news, and know—judging by ourselves, and representing to ourselves what we should feel in like circumstances—in what manner their sensibility will be touched, or what emotions they will naturally experience.

CHAPTER IV.

DREAMS.

THE automatic activity of the cerebral cells reveals itself also, in a very distinct manner, at night in the form of persistent impressions—dreams. It naturally follows, from what we have already explained, that, in reality, dreams are nothing but the persistent vibration of certain groups of cells in a condition of erethism, when the greater number of their fellows are already plunged into the collapse of sleep.

This persistent vibration of the nervous elements may be explained physiologically, either by the fact of a strong super-excitation occurring in consequence of too prolonged exercise, or because of some special excitability, some peculiar receptive condition of certain cell-territories, which have felt external stimulations more intensely than the neighbouring regions. It is, then, sufficient that a certain number of them shall continue in vibration, in order that these shall become centres of appeal for other agglomerations of cells with which they have either more intimate affinities, or more or less facility of anastomosis. Hence arises a series of revivals of past impressions, of which we scarcely catch the sense, but which have secret connexions one with another (unconscious memory); a series of unexpected and disorderly ideas, which follow one another in the strangest forms. They are developed by the mere automatic forces of the cerebral cells abandoned to their own will, and freed from the directing influence of sensorial impressions (visual impressions), which, in the natural order of things, keep them awake and regulate their diurnal mode of activity.* Hence those unexpected apparitions which surprise us in dreams, and which are nothing but the result of the partial awakening of certain cells, which thus cause a series of long-forgotten impressions to rise again in the *sensorium*. These, however, are in reality never anything but impressions which make a part of the stores we have acquired, which reveal themselves in our dreams, and which, probably under the influence of local conditions of circulation, neighbouring impressions, &c., revive from out the depths of our past. To dream of anything, we must have seen it in one fashion or another. It is not rare, in seeking out the origin of certain dreams, to recognize that a great number have a more or less direct relation to an impression that was more or less strongly impressed upon us in the waking state, and that they are but a species of echo of this impression, associated with more or less heterogeneous impressions.†

* The direct influence of the arrival of sensorial influences, and visual in particular, on the regularity of the play of the cerebral cells, is such that in a patient, whose case is reported by Baillarger, in a waking condition it was sufficient to lower his eyelids and thus suppress the arrival of optic impressions in his brain, to cause apparitions of various objects, of which he had previously no idea, to appear to him immediately; and this by means of the mere automatic forces of the brain, which resumed their course, *motu proprio*, when external impressions were excluded. ("Annales Médico-psychol.," vol. vi. p. 178.)

† A young girl, mentioned by Prus, believed she had merited eternal punishment by having given way to too tender a sentiment. This idea preoccupied

Hence, again, those curious phenomena through which dreams produce in us subsequent emotions which so profoundly overwhelm us.

These emotions, as we have said, are necessarily associated with the former impressions which have given them birth. They live with the same life, so that the appeal of the former inevitably evokes its fellow. If the first sight of any person, or spectacle, or scene, have caused us a moment of pleasure or anxiety, the reminiscence evoked by the same objects will be followed by the same emotions of our natural sensibility. In the domain of dreams the same phenomena unfold themselves in the same concatenation; if one idea or agreeable memory arise in the psychical sphere, in consequence of a state of erethism in a special region of the brain, immediately an analogous condition of concomitant satisfaction will be felt in the *sensorium*—if an idea of quite another nature should arise, either spontaneously or through some disturbance occurring in the visceral innervation (cardiac anxiety, gastric pain, irritation of the genital organs); if the mind, for instance, gives birth to conceptions regarding precipices, scenes of murder, etc.; at the same time analogous states are developed in the emotional regions of our organism, and this artificial evocation of sensibility may produce a shock—a dynamic effect—intense and powerful enough to awaken the sleeping cerebral cells. Thus it is always the mere automatic forces of the nervous elements which regulate and govern the world of

---

her vividly for some time, until she believed one night that she saw and heard a messenger from heaven who announced her eternal damnation and that of her family. ("Annales Médico-psycholog.," vol. iii. p. 103.)

our thoughts and sentiments, sleeping as well a waking.

This invincible persistence of the activity of the cerebral cells reveals itself very clearly in the following circumstances. Just as the phenomena of their diurnal encroach upon those of their nocturnal activity, so, inversely, we often see those of their nocturnal activity perpetuated while we lie awake or are dreaming, and not ceasing even at the moment of total awaking of the brain.

We all know that when we awake we may, for a while, generally in the early morning, preserve the remembrance of the dreams that have traversed our brain during the night. We have all met with a certain number of persons who are more or less intensely pre-occupied with the dreams that have assailed them. Certain feeble characters are even more or less depressed by them, and preserve a painful impression, which they sometimes consider a true presentiment of what may happen to them.

It is, however, in certain forms of mental maladies that this power of dreams to persist during waking moments acquires the greatest degree of intensity. Thus we see a certain number of patients, paralytics, or victims of hallucination, change the character of their delirious conceptions and take up new ideas, which are nothing but persistent dreams that have been suddenly developed in their minds at night. Thus when I find patients expressing certain ideas and certain apprehensions in the morning, I infer that they have had terrifying impressions of a certain kind, and should expect to see this emotional condition persist in the form in which it was implanted until it became a fixed and permanent idea, a delirious conception of new formation. We must, therefore, seek the origin of the transformation of certain deliriums, and certain ideas which unexpectedly change the direction of the mental state of certain individuals, in the continuation of the period of erethism of certain groups of cerebral cells, which have come into action during the period of sleep, and continue to vibrate even during the waking condition.

CHAPTER V.

DEVELOPMENT OF AUTOMATIC ACTIVITY.

THE automatic activity of the nervous elements awakes at variable epochs, according to the precocity of morphological development of these elements. Thus, the spinal axis being more rapidly developed than the brain, in the regular evolution of the nervous system, automatic manifestations may take place in it, even while they are very imperfectly developed in the cerebral grey substance.

The manifestations of automatic activity in the brain follow by degrees the progress of physical development, and this in a very rapid fashion. When the arrival of external excitations calls its sensibility into play in the different foci in which it is concentrated, the child begins to benefit by the automatic development of the activity of his cerebral instrument, and it is by this process of continual absorption of impressions, which incessantly reverberate in all the regions of the *sensorium*, that the external world penetrates into him, and that by the silent evolution of the specific energies of the cerebral cells, his mental development goes on with that prodigious rapidity which excites our astonishment.

We have all witnessed and wondered at the incessant mobility of these young creatures, and their insatiable appetite for knowledge of the external world. They want to embrace it with their little hands, to touch all that surrounds them, and take cognizance of everything that comes within their ken. We have been struck by their deep reflections, of which the scope, the logic, and the subtlety, surprise us all the more from their being exempt from all reserve, and their coming to light simply from the natural play of the cerebral activity abandoned to its frankest manifestations.

Moreover, while the physical world penetrates into him and leaves its traces upon him, the child begins to feel emotions, and have developed in him the primordial elements of common sensibility. He feels very distinctly, of his own accord and by the mere energies of the vitality of his own *sensorium*, what things and persons are agreeable or disagreeable to him.

It is through the action of the same unconscious vital forces that his first sentiments arise, and are developed and perpetuated; by means of a blind original force, without the direct intervention of the personality.

Childhood and youth are the two phases of life during which the automatic activity of the cerebral elements reveals itself with the greatest energy. It is the period when memory has the greatest vigour, when the sensibility of the cerebral cell is most exquisite, either to feel the excitations which thrill through it, or to retain them. It is also that at which its reactionary faculties are most intense.

It is indeed the period when ideas are associated with the greatest rapidity, when the conjunction of new with old ideas takes place instantaneously, when individual spontaneity and personal originality burst forth in the most pronounced manner, and when, in fact, the man appears with the cerebral temperament which specifically characterizes him.

As maturity approaches, the automatic energies of the cells become gradually less intense. Their sensibility is already dulled in consequence of the multitude of impressions which have affected them in turn; their appetite for unknown things is less intense; it is their period of beginning saturation. The thirst for knowing and registering new facts calms down by degrees, and the mental forces then concentrate themselves for the regular classification of acquired riches, the methodic grouping of facts belonging to the past, and the calling into activity of the materials long ago accumulated, which serve for the building up of our judgments, the formation of our thoughts, the maturation of our reflections—so that if the human brain has already lost something of its freshness, and the juvenility of its manner of feeling, it has gained, *per contra*, the fruits of acquired experience. It knows and automatically expresses what it knows; and these different modes in which the human personality reveals itself as regards its external manifestations, represent the true synthesis of all the mental activities in their full expansion.*

* It is curious to observe practically, in every-day life, how variable the degree of the automatic energies is in different individuals, as regards the rapidity of the transmission of nervous excitations to the brain and of the consecutive reactions. We know indeed how many individuals there are who, as it is said, have their understandings slow, sluggish, and hardly permeable by the stimulations of the external world, which are radiated towards them. Every one knows that a great number of people exist, who, although very intelligent as regards a certain class of ideas, are incapable of appreciating the

The effects of progressive senility are marked by insensible gradations in human brains, by a slow and gradual enfeebling of the automatic activity of their elements. That failure of appetite for, and curiosity about, new things, which is already marked in the preceding stage, becomes more and more distinct. That deadening of the sensibility, which expresses the complete saturation of the elements of the *sensorium* and their incapacity for maintaining a condition of erethism, takes more and more significant forms. The human brain experiences the need of prolonged repose; the ardour of the grand struggle for life becomes painful to it. A retreat is sounded from a great number of social careers. Thus it is that that period of inactivity which inevitably awaits each individual, as regards the social part he has played, physiologically expresses the slow and gradual wearing out of the energies of automatic life, which by degrees cease to vibrate, and betray by their slackening the progressive dulling of the sensibility of the cerebral cell.

In proportion, then, as sensibility grows languid, and the faculty of erethism loses its energy in the elements of the *sensorium*, the external manifestations of the life of the brain undergo a parallel retrogressive movement. Repose and silence insensibly invade them. The field of the ideas and sentiments grows narrower; intellectual spontaneity becomes languid, and verbal expression, and

association of two incongruous ones. They but very slowly comprehend facetiæ and plays upon words, and in conversation, when a play upon words has been made and every one has laughed, they alone hang fire, and show by a tardy burst of laughter that the hilarious effect has at last been produced upon them also.

conversation, dried up at the fountain-head, cease to be interesting and endowed with a spontaneous character. The man who has nothing to say, who has but a few notes of his personality to set vibrating, speaks no more or says but little, at least if we do not take for original conversation those vapid phrases that men think themselves obliged to exchange, when they are in each other's company, and of which the inanity, to some extent *reflex*, only covers the absence of ideas and sentiments.

Thus it is that, through the necessary connections which unite all the zones of cerebral activity, the manifestations of senility by degrees gain ground in the psycho-intellectual spheres. The mere fact that there are regions of the brain which have primarily been struck with stupor and histological degeneration, causes the same retrogressive processes to radiate to a distance, and, through secondary lesions, inevitably to produce the symptoms of senility and more or less progressive dementia.

CHAPTER VI.

FUNCTIONAL PERTURBATIONS OF AUTOMATIC ACTIVITY.

IT is in the series of morbid phenomena peculiar to mental diseases that the processes of automatic activity generally present themselves with their most significant characters of intensity, and in the most diverse forms.

It is, in fact, the automatic activity of the cerebral cell that always more or less comes into play, in general or partial delirium, and in irresistible impulse, being everywhere essentially active, and everywhere present. It is always this that reveals itself with those characters of irresistibility, and that evident freedom from voluntary action which are its special characteristics.

Thus general delirium, with that exuberance of thoughts which clash and associate in the most unexpected manner in the brain of the patient, is the highest expression of the automatic activity of the cerebral cells in a condition of irrepressible erethism. It is enough to have seen patients at this period of extreme over-excitement, to recognize the fact that the will is powerless to repress the disorder; that the very elements that constitute human personality are themselves in disarray; and that in this agitation these incoherent words, these sonorous explosions to which all the cerebral elements contribute in so unconscious a manner, we cannot fail to recognize the tumultuous expression of the forces of normal energy, unchained and hurried into a very whirlwind of morbid over-activity.

In some forms of partial delirium, we see patients less vehemently hurried along in spite of themselves; incessantly delirious on certain points, conceiving the same delirious conceptions, always repeating the same phrases, without perceiving that their ideas are in complete discord with reality. Thus they say they are ruined, robbed by every one, poisoned; and even if anyone should reason with them, proof in hand, respecting the folly of their apprehensions, and reassure them in a thousand ways, the automatic activities of their brain are so set in a false direction that they incessantly return to it, just as a contracted member on being extended will resume its former position. They are perpetually complaining, they incessantly repeat the same phrases, the same vague apprehensions, and unconsciously fall back into the same ruts followed without conviction, without participation of their conscious personality, merely by dint of the automatic forces of their troubled mind.*

In other circumstances automatic activity is exercised in a morbid manner within a comparatively limited circle, and only engages certain zones of the cortical substance, the others remaining comparatively unaffected; as we see for instance, certain cutaneous

---

* See Billod, "Annales Médico-psychol.," 1861, p. 541. Lesions of Association of ideas; fixed ideas.

phenomena reveal themselves in patches, in little islets on the surface of the skin, leaving sound regions at intervals. Thus, in the cases to which we allude, the perceptive regions of the *sensorium*—those in which the manifestations of conscious personality take place, are sometimes spared, and in a condition of complete integrity, while the neighbouring regions are invaded by different kinds of morbid processes ; and then we witness a strange phenomenon—a sort of duplication of the mental unity. The individual, thus divided into two parts—one portion of himself remaining healthy, while the other is at the mercy of the phenomena of automatic, involuntary impulse—looks on, as *a conscious spectator*, at certain extravagant acts that he is forced to commit, at certain senseless words that he utters. He is in a manner reduced to the painful position of the tetanic patient, who at the moment of the attack sees his muscles escape from the influence of his will, contract under the influence of the cells of the spinal cord, in a paroxysm of automatic, irresistible activity, and thus become unwieldable instruments which cease to belong to him.

The annals of mental diseases include numerous examples of this state of dissociation of the vital forces of cerebral activity. There are patients sometimes who write and describe their distresses—the involuntary agonies through which they pass, the words they have pronounced unwittingly ; how they are impelled to speak in spite of themselves, to say what they would not have wished to say, to go through ridiculous gesticulations, and to commit extravagances they believe themselves incapable of restraining.

A lady described by Falret uttered cries, committed all sorts of disorderly acts, and felt herself the more to be pitied, because she knew that they were acts of madness, but could not avoid committing them.* A patient described by Moreau (of Tours) presented analogous symptoms:—

X——, in consequence of grief, became irritable in temper, and was seized with eccentric ideas which his reason disapproved. Suddenly the idea of tossing his bed would occur to him; but he would ask himself what was the good of it. Or he would be tempted to throw his hat upon the ground without a motive. In conversation if any one dared to contradict him, a sudden desire to beat his adversary seized him, but he restrained himself by thinking of the absurdity he would thus commit; and a crowd of delirious ideas would incessantly traverse his mind without his permitting any one to suspect him of madness, so short was the duration of his paroxysm.†

These strange phenomena, these general or partial deliriums, these strange impulses, of which we see abortive specimens in certain pregnant women, constitute, in the form of suicidal or homicidal impulses, the essential morbid elements, and in a manner the primary factors of mental pathology. They are all, in different degrees, derived from the fundamental properties of the cerebral cell, from its automatic activity past into a phase of inveterate erethism. It is always the same fundamental property which is at the bottom of all morbid

---

* Falret, "Annales Médico-psycholog.," 1870, p. 117. Conscious Lunatics.
† Unusual impulses, with disorder of intelligence. "Annales Médico-psychol.," p. 84, 1857.

manifestations of the brain, and which, always present, always identical with itself, either in normal or morbid conditions of the life of the brain, becomes the source of all the disorders and all the anomalies of mental life.

PART III.

EVOLUTION OF THE PROCESSES OF CEREBRAL ACTIVITY.

HAVING thus far considered the elements of cerebral activity as individual simple forces in the statical condition, we shall, in this third part of the work, consider them from a dynamic point of view, as living forces in movement, in combination one with another, effecting reciprocal reactions, and co-operating in the different modes of mental activity.

One general fact governs the essential organization of the cerebral cortex (see p. 15). This fact is the admirable order, the regular subordination which is established in the grouping and methodical distribution of all the elements of this cortical substance. In all its regions the zones of cells are arranged one below another in thicker or thinner strata; they are strictly united one with another, both transversely and horizontally as regards this substance; the regions of small cells, moreover, everywhere occupy the superficial sub-meningeal zones, while the regions of large cells are localized in the deep regions, and communicate with the preceding by a series of intermediate links—strata of cells which serve as a transition between these two isolated regions. If we compare this simple disposition, which is the anatomic formula in which the economy of the constitution of the cerebral cortex is epitomized, with that which regulates the reciprocal relations of the nerve-cells of the spinal cord, we immediately perceive that it presents certain analogical characters which in a manner explain themselves; and we cannot avoid recognizing the fact that if in both instances there are analogies from an anatomic point of view, that is to say regions of large cells indirectly anastomosing one with another, there should similarly be physiological analogies as regards the mechanism of the activity of these similar elements.

Now, as experience proves that the nervous currents pass across the spinal cord from the smaller to the larger cells, and that these latter never enter into activity spontaneously, but merely in consequence of an incidental excito-motor excitation, which they simply reflect, we cannot help admitting, from the most legitimate analogy, that the nervous actions must be evolved in a similar manner throughout the stratified elements of the zones of the cerebral cortex. We may therefore conclude that the regions of small cells in the cortex represent in the brain the posterior grey regions of the spinal cord, and that, like them, they are the territory of dissemination of sensitive impressions, designed to retain them, store them up, and afterwards propagate them to the subjacent zones.

From the clear analogies which exist between these two spheres of nervous activity, the spinal cord and the brain, we are therefore led to the conclusion that the different zones of the cortical substance, taken as a whole, represent, as it were, a series of sensori-motor organs conceived on the same plan as that of the similar organs of the spinal axis; that the nervous activities are developed throughout its tissue as throughout that of the spinal grey matter; and that in both instances the processes which take place are always —except for differences of medium, the different qualities of the elements called into play, the amplitude and complexity of the different phases of which they are composed—similar processes, reducible to the same primordial phenomena. It is always a phenomenon of sensibility that produces the movement, and excites the activity of the motor cell; and the motor act itself, whether we have to do with the spinal cord or the brain, is always, as regards its dynamic signification, merely a secondary and subordinate phenomenon, the return effect of a sensitive impression transformed.

This being the case, the phenomena of cerebral activity, as regards their successive development, may be briefly reduced to a series of processes—of regularly-linked physiological operations, all derived one from another, becoming complicated in their diverse phases, but always having a common basis of elementary operations.

It is always a phenomenon of sensibility, an anterior sensorial impression, present or past, that marks the point of departure, and becomes, in a more or less sensible form, the primary stimulation that induces the movement. In a word, it is always an agitation of the *sensorium*, an emotion of the personality, that expresses, through the infinite series of cerebral operations, the condition of erethism which it has experienced.

Hence there are three natural phases under which we shall successively consider the mode of evolution of the different processes of cerebral activity.

1. A phase of incidence, which corresponds to the moment when the external impressions arrive in the plexuses of the *sensorium* and are perceived there (phenomenon of attention—genesis of the notion of personality—conscious perception).

2. An intermediate phase, during which the affected elements of the cortical substance enter into active participation with the external impression, transformed into a psycho-intellectual excitation. (Dissemination of sensorial impressions in the psycho-intellectual sphere —evolution and transformation of these impressions— operations of the judgment, etc.)

3. A phase of reflexion, which corresponds to the moment in which the primordial excitation, being propagated through the plexuses of the cortex, passes outwards, and expresses, by voluntary motor reactions, the different states of the previously impressed *sensorium*. (Genesis and evolution of the phenomena of voluntary motion.)

BOOK I.

PHASE OF INCIDENCE OF THE PROCESSES OF CEREBRAL ACTIVITY.

CHAPTER I.

ATTENTION.

THE period of incidence of the process of cerebral activity occurs at the moment when the sensorial excitations darted from the different centres of the optic thalami are distributed to the different regions of the *sensorium*, upon which they thus produce a consecutive impression (Fig. 6, p. 61). We have already several times insisted upon the different phases of evolution of the phenomena of sensibility, and shown that this simple physical impression produced by the external world is transformed, as it becomes incorporated with the organic tissues, into nervous vibrations, and that these nervous vibrations, passing through successive agglomerations of cells, undergo the action of the different media through which they pass, until they arrive transformed and purified in the plexuses of the cortical substance, which are set in motion, impressed, and vivified by them alone.

The regions of the *sensorium*, which are the living sources that feed all the activities of animal life as at a common reservoir, are then, before they react by radiating outwards the forces that they create on the spot, themselves the tributaries of excitations from the external world, which, like an electric spark dispersed throughout their tissues, suddenly excite and develop their latent energies. It is necessary, therefore, as a fundamental condition of the evolution of the intracerebral processes, that sensorial impressions shall be regularly conducted during their period of incidence, that they shall be distributed according to the physiological laws we have described, and that, besides, they shall be *received, propagated, and retained*. At this precise moment of cerebral activity, a delicate, precise, and rapid phenomenon takes place. This is called the *phenomenon of attention*. It is quite comparable to that which we have already described at the other pole of the nervous system, at the moment when sensitive impressions come into contact with the peripheral plexuses, and when the external excitation, becoming incorporated with the nervous tissue, loses in an instant the qualities of a purely physical, to assume those of a purely nervous excitation.

At the periphery, at the precise moment when the external excitation, represented either by a luminous or a sonorous vibration, or by a material impression, impinges upon the sensorial plexuses, an inward phenomenon of impregnation or transformation of force occurs. The natural sensibility of the nervous element is affected: it becomes *erect*, is arrested, is *attentive;* and from this intimate contact with the external vibration

ATTENTION.

it enters into a new state; a specific impression is made upon it, which passes from the external world from which it is derived, to explode in the *sensorium* itself.

The plexuses of the *sensorium*, which themselves represent a vast sensitive surface open to external excitations, are the theatre of phenomena of the same kind. For there each excitation from the external world arrives in a quintessential form, intellectualized by the metabolic action of the centres of the optic thalamus. Henceforth it represents only the distant and transformed echo of an impression, which was purely physical when it made its first appearance in the organism. Here also, in order that this incident impression shall penetrate into the plexuses of the *sensorium* and become incorporated with them, it is necessary that it shall find in them proper conditions of receptivity, that their natural sensibility shall be excited, that it shall be seized upon, and that a species of similar erethism shall be developed. This is what, in fact, takes place at the moment when the excitation arrives in the *sensorium*. Its impregnation does not take place *coldly*, nor without a local reaction and an active participation of the nervous element thrown into agitation. There is a period of physiological erethism which this element then manifests at a given and variable point in the cerebral cortex. It is, in fact, actually recognized that, at the moment when this subtle phenomenon takes place, there is a local development of heat, which is disengaged in the cerebral region that becomes active (experiments of Schiff, see p. 77), and that this reaction expresses the active participation, the *attentive state* of the elements of the *sensorium* which receive the excitation, at the moment when they are impregnated by it, and transform the purely sensorial excitation into a psychical impression.

Attention, which marks the first phase of all the processes of cerebral activity, is, then, a phenomenon similar to all those developed in the peripheral plexuses of the system when they are impressed by excitations from the external world. It is the *sensorium* itself, the sensitive plexuses of our organism, the *conceptive* regions wherein the notion of our personality dwells, that are immediately engaged and become *conscious* of the inward phenomenon which occurs. It is from this very fact that the operations of the attention are always, *par excellence, conscious* operations, which imply the necessary participation of the entire human personality.

Thus, then, in order that the processes of cerebral activity, by virtue of which attention is exercised, shall be evolved in a regular manner, it is necessary that two indispensable conditions shall concur: on the one hand the registration of the first sensorial impression, regularly effected in the peripheral plexuses at the moment of its genesis; and on the other hand the active, spontaneous, and original participation of the elements of the *sensorium*, which must vibrate in a concordant manner and enter into unison with the impressions radiated from the peripheral regions. It is necessary, then, that between these two poles of the system, there shall be a simultaneous co-operation.

On the other hand, it is also necessary that at the moment when the excitation from the external world arrives in the *sensorium*, it shall be introduced methodically, and in a gradual manner; that it shall work its passage independently; and that, at the moment at which it is there deposited, it shall vibrate alone, and alone imprint the records of its presence upon the plexuses of the *sensorium*. In a word, it is necessary even in optical experiments, when we wish to study the elementary properties of a luminous ray, that we shall carefully eliminate the rays of diffused light, and cover the head with a black veil to eliminate from the eye the incident rays—and just so, for the perfect accomplishment of the phenomena of *conscious* attention, in order that these shall produce their maximum of effect it is necessary that simultaneous and approximate impressions shall not come to join the principal impression, and eclipse by their presence its intra-cerebral radiation. To be attentive it is, then, necessary simultaneously to receive impressions from without, and to admit them only in a gradual and successive manner. Without these fundamental conditions the process is abortive, and confusion of impressions and want of precision in the notions acquired, are the sole and ultimate result of this abortive operation.

It is necessary, then, that one single impression at a time shall be imprinted upon the *sensorium*, and that, moreover, the elements of the *sensorium* shall themselves be in a kind of silence and relative calm.

In fact, where lively preoccupations, or a prolonged intellectual effort maintain in certain zones of the cerebral substance a period of erethism more or less persistent, the result is that this local over-activity, simply from its being in possession of the field where it has originated, will stifle by its intensity impressions from the external world. Subjectivity predominates, and itself alone absorbs the cerebral activity; so that external impressions grow dull on arriving, only penetrating into the regions of conscious personality, when forestalled by excitations originating on the spot. They are consequently as though they had never arrived. It is necessary, therefore, that at the moment when the external impression arrives in the brain, it shall find the sensitive regions available, in a healthy condition, and free from every local cause of internal excitation.

In order that the process shall be completely effected, another special condition is finally necessary—a condition of receptivity similar to that which, as we have seen, must exist in the peripheral regions of the system. It is necessary that the impressed cerebral cell shall, like the cell of the sensorial plexuses, be endowed with a certain special retentive power, and with a certain energy for supporting fatigue; for it is at the expense of its substance that it produces movement, vibrates, enters into erethism, and becomes attentive.*

We all know that it is impossible to prolong efforts of attention beyond certain limits,—that we cannot, for instance, fix our attention in an undetermined manner for a prolonged period upon some petty fact, which only involves a single sensorial impression. It is only by mnemonic artifices that we succeed in reviving the fleeting impression by a series of successive instigations producing continuance of the

* The phenomena of fatigue, or functional exhaustion in the nervous elements, reveal themselves very clearly, as we have already explained in speaking of the retina, which is rapidly fatigued by certain luminous rays which have too continuously affected it.

act of attention.* This *distraction* takes place by reason of the vital forces of cerebral activity themselves; for as certain regions of the brain, fatigued by sustained attention become inactive, other cell territories, reposed and fresh, automatically come into action, by virtue of their native energies, and monopolize the vital forces of these regions of the conscious personality for their own profit. Thus, we may say that the failure of attention, and easy distraction, imply the rapid fatigue or need of repose of the cerebral cells; so that we are led to the conclusion that the vigour of attention is, to some extent, a measure of the degree of vigour of the mental faculties; that it is the external expression of the energy and vitality of the cerebral elements; as, in the appreciation of motor phenomena, the continuity of effort is proportional to the disposable motor force.†

Functional Perturbations.—The processes of attention represent, then, as we have just explained, a synthesis of the active operations of the brain, in which the phenomena which occur in the periphery of the nervous system in the sensorial regions on the one hand, and the phenomena which are developed in the central regions of the *sensorium* where they come directly in contact with external excitation, on the other hand, are fused together. We can therefore comprehend how, when these two fundamental conditions happen to be disturbed

* To make use of an analogous illustration, we know, that the continuity of muscular contraction is merely the result of a series of successive shocks.

† Hemiplegic patients, whose brains are partially disorganized, are very rapidly fatigued as regards the amount of attention they lend those who speak to them. Patients with dementia, whose cortical substance is more or less profoundly degenerated, are in the same condition: they only lend a very limited degree of attention to words addressed to them, and cannot sustain a connected conversation for more than a very few minutes.

in their constituent elements, whether in the peripheral or central regions, the processes of attention are, at the same time, disturbed and arrested in their regular evolution.

Thus, when it is the peripheral regions that cease to be in their normal conditions of receptivity, when the sensorial apparatuses are not adjusted in the required direction—when, for instance, certain sensorial plexuses are struck with anæsthesia—the unperceived and unregistered excitations from the external world are practically absent as far as the sensorium is concerned.

Thus, physicians are well aware how indifferent all anæsthetic individuals are to oscillations of temperature in the atmosphere in contact with their bodies; how little *attention* they give to what directly touches them, yet only produces in them a confused impression; how certain individuals with well-marked myopia have a vague and blinking mode of looking at those things in the surrounding visual field which they do not see, and to which consequently they pay no attention; how easily the deaf are distracted, only following with much trouble the series of ideas brought before them; how in a great number of individuals attacked with mental diseases the systematization of certain forms of delirium has no other cause than a sympathetic irritation, or sensitive disturbances radiating from the peripheral regions which alone attract their attention. (Phenomena of hypochondria.) We know also, that when these same regions are excited, and have arrived at the pitch of pain, they keep the faculties of *attention* in the *sensorium* in a permanent condition of erethism. Every one, indeed, knows how vehement a reaction the painful spot has upon the *sensorium* when we suffer in any point whatsoever of our sensitive territory; how completely it absorbs all our attention; and how profoundly its painful radiation jars upon our conscious personality, which is forced to pay unbroken attention to what is occurring.

In other circumstances it is the central regions that are engaged, and therefore place an obstacle in the way of the regular perfection of the processes of attention.

Thus, in idiots and imbeciles, the state of imperfection of the nervous system, either of the peripheral or central regions, renders them dull in perceiving, regularly, impressions from without. Their senses are dulled, their sensibility obtuse, and thus they are capable of but a slight degree of attention. They see badly, hear badly, feel badly, and their *sensorium* is in consequence in a similar condition of sensitive poverty. Its impressionability for the things of the external world is at a minimum, its sensibility weak, and consequently it is difficult to provoke the condition of physiological erethism necessary for the absorption of the external impression.

Thus it is that defect of attention is the rule in these special forms of mental degradation, and it is not without reason that Esquirol has connected the inaptitude of idiots for education with their defect of attention.*

---

* "Imbeciles and idiots are deprived of the faculty of attention," says Esquirol. "I have repeatedly made this observation with regard to them. Wishing to mould in plaster a great number of insane persons, I was able to do so even with furious maniacs and melancholics; but I never could succeed in getting imbeciles to keep their eyes closed long enough to apply the plaster, however good the intention with which they went to work. I have even

In all forms of mental disease the faculty of attention becomes gradually weaker, and presents, according to the intensity of the morbid process, different and fatally progressive modifications.

In a general way, in persons with hallucinations, individuals attacked with acute or chronic mania, etc. etc., the forces of attention cease to take effect, the phenomena of the external world no longer produce in the *sensorium* anything more than an abortive impression. Morbid excitations are developed locally in the very regions of subjectivity, which become erethised of their own accord, and thus virtually become an insurmountable barrier between the individual and the surrounding medium. The patient, thus shut up from external sounds, a stranger to everything that passes around him, lends but an *inattentive* ear to the things of the external world. He lives, as people say, in himself, upon remembrances of the past, and upon his habitual delirious conceptions. Days pass away, the world goes by, events succeed around him, he no longer pays any attention, and the progressive indifference and invading apathy which manifest themselves in him, attest the gradual exhaustion of the vital forces of his mental activity.*

seen them cry because the mould of their heads has not succeeded, and several times vainly make the attempt to keep the pose that was given them, not being able to keep their eyes shut more than a minute or two." (Esquirol, Tome I. p. 11.)

* There are circumstances in which, in the case of insane patients whose intellectual faculties are not as yet quite extinguished, we see certain sharp and unexpected external excitations come in to produce a happy modification of their mental condition and provoke in them some manifestations of attention. Thus Vigna has reported the history of certain individuals, who though apparently incapable of the simplest reasoning, when brought into the presence of a person who overawed them, a magistrate for instance, were much excited by the influence of the new circumstances in which they were placed, and then

produced the elements of a regular defence, and thus succeeded in preventing a judgment of interdiction. ("Annales Médico-psychol." 1871, p. 17.)

Baillarger has similarly noticed that in certain patients with hallucinations, vivid and sudden impressions may arrest the morbid working of their brains and induce attention to what is going on around them. "At the moment of the arrival of the physician," he says, "hallucinations disappear. They cease to hear voices; but one has scarcely left them before they fall back into their false conceptions." (Annales "Médico-psychol." 1845, vol. vi. p. 185.)

CHAPTER II.

CONSTITUTION OF THE SPHERE OF PSYCHO-INTELLECTUAL ACTIVITY.

WHEN once the external excitation is disseminated through the plexuses of the cortical substance, and incorporated in the *sensorium*, developing in it the specific energies of the cerebral cells that have received it, this new medium itself comes into play, and reacts in the direct line of its latent capacities.

The sphere of psycho-intellectual activity then exhibits all its natural riches, all the stores of its awakened sensibility. It is suddenly thrown into agitation, reacts, and thus develops the marvellous capacities with which it is fundamentally endowed. This new medium which comes into play, comprehends, as we have said, the sum of the purely psychical and purely intellectual phenomena of the living organism. It is the *regio princeps* of the organism, in which all ends, from which all begins, and which is the epitome of the vital forces of mental activity.

Now, how is this double sphere of activity, which from the dynamic point of view presents characters so distinctly marked, and yet so intimately fused together, constituted? How may it be ideally conceived as regards the cortical structure?

These are questions to which it is at present impossible to give a completely satisfactory reply. We merely know, from the anatomical data we have already laid before the reader, that the cerebral cortex is made up of a series of cells superposed in independent zones and yet united one with another; and that these plexuses of cells directly receive external excitations, chiefly within a specially circumscribed region, thus forming a vast surface of reception for these excitations, and in the strict acceptation of the word a true *sensorium commune*.

Now these plexuses of the *sensorium*, constituted by the different submeningeal zones of cortical cells, are not merely inert screens, nervous zones destined to receive and passively register the images of the external world. They are sensitive, living, emotional plexuses, which become erect in a peculiar manner in presence of the stimulating excitation, and which, on its arrival, like their fellows in the peripheral regions, give evidence of the various manners in which they may be impressed. They live, they feel, and what is more, they remember; for then it is that this new property of preserving records of past impressions, appearing in full force, gives a special character of permanence to all the excitations that arrive, and enables them to survive themselves, to prolong their existence in the form of memories, and to be marked in the calendar of our sensitive impressions with a special co-efficient of pleasure or pain.

Thus by these two fundamental conditions, the arrival of the external excitation, and the appropriate reactions of the cerebral medium in which it is received, new forces are created in the brain, a special sphere of nervous activity is developed, in which the natural sensibility of our being, our conscious personality, represented with all its elements (see p. 105) in the tissue of the *sensorium*, comes to life, expands, and perfects itself, by the coming into play of the natural sensibility of the elements which compose it. Thus, consequently, former impressions are inevitably associated with recent ones; the past life of diverse emotions, the memory of days of joy or sorrow, is incessantly in contact with the conceptive regions of mental activity; and in fine, if the sphere of purely psychical activity be considered from a dynamic point of view, as the resultant of all the impressions of our sensibility present and past, associated with the events of our current life; from an anatomical point of view, it may be conceived of as localized in all that series of nervous elements which constitute the plexuses of the *sensorium*.

On the other hand, if, up to a certain point, we have some precise data which permit us to suppose that certain regions of the cortex (the regions of the small cells) play the part of a common reservoir as regards the external impressions which are distributed among them, and consequently become the special territory of the phenomena of mental sensibility, it is as yet almost entirely impossible to obtain any precise data as to the real constitution and topographic situation of the field of intellectual activity proper. It is only artificially, and in a roundabout way, that we can succeed in grouping a few facts with regard to the subject.

The study of mental diseases shows us, indeed, in a precise manner, that in a great number of cases the regions of intellectual activity may be spared when the purely emotional regions, the regions of the *sensorium*, are profoundly disturbed. We see a great number of insane patients affected with melancholia, groaning over their fate, over the persecutions to which they are subjected, mourning incessantly about trifles, and yet capable of taking note of what is going on around them, and, in the midst of the disorders of their agitated sensibility, discerning perfectly what happens to them, and sometimes making very just reflections.

This dissociation of the purely emotional and purely intellectual regions, which may be unequally affected, proves, then, distinctly the complete functional independence of the intellectual sphere proper, and that of mental emotivity and sensibility.

Now, where is the seat of this sphere of intellectual activity, which has its own special domain, its peculiar autonomy in the midst of the operations of the brain ; and what are its connections with the different groups of cells in the cortex?

Here again, up to the present time, we have only conjectures and probabilities to offer.

In taking note, however, of the order and progress of those processes of cerebral activity which spread by propagation from the superficial submeningeal regions to the deeper regions of the cortex, we cannot help admitting that the sphere of intellectual activity can only be set in motion secondarily and consecutively to the impression of the plexuses of the *sensorium*. The plexuses of the *sensorium* receive the first onset of the external excitations, and sift them to some extent before propagating them to the subjacent zones. They are the natural frontiers by which all the excitations of the external world must necessarily pass. Now, this natural frontier topographically occupies the superficial regions of the cortex; we may, therefore, provisionally admit that the zones of cells subjacent to the plexuses of the *sensorium*, with which they are continuous, are those which, without its being possible precisely to limit their thickness, may be considered the field of action of the operations of the intellect proper.

This theory—which squares with the facts of daily observation, which show us every moment how closely connected the activity of the intellect is with that of the senses, and that the intellect, in order to come into play, must first have received its stimulation from the external world—explains, at the same time, how it is that the phenomena of intellectual activity, from the very fact that they are exercised in an isolated territory of the cortical substance, are apt to show themselves in an automatic manner, under a special aspect, as a completely independent sphere of activity.

However it may be, the intellectual sphere, considered in itself, participates in the same dynamic manifestations we have observed as regards its fellow, the psychical sphere.

Like this, it becomes active under the influence of the excitations of the external world, which beget in it movement and life; like it also, it becomes erect in consequence, and develops its natural energies. But here sensibility and emotivity no longer play the first part, as when the plexuses of the *sensorium* proper are in agitation; the purely automatic activities of the cells then develop themselves with a specific energy which is very significant. If sensibility be the dominant note of the psychical activities, automatism is the characteristic of this special field of the life of the brain.

Everything, in fact, records itself automatically, and outside of the will. Without our knowledge certain ideas present themselves, certain associations are effected among themselves, certain reminiscences are evoked. Everything in this special domain is done in an irresistible, inevitable, unconscious manner, by means of the automatic activity which reigns as sovereign and governs the series of the operations of the intellect. It is it, indeed, that creates new relations, stores up our memories, and daily tacks them on to more recent events. It is always present, always active, and, by a strange phenomenon of which we are incessantly the dupes, it comes to light in the form of *spontaneity* in our ideas, our words, our acts, thus becoming, as we have already indicated, the most living expression of the freshness and vitality of the cerebral regions which have given it birth.

Thus, then, the sphere of psychical and that of intellectual activity represent, isolatedly, each from the point of view of its dynamic action, the most complete epitome of the fundamental properties of the nervous matter. In the first, the phenomena of sensibility, with all that is most exquisite and most perfect in them, predominate; in the second, the phenomena of automatic life. These two regions of cerebral activity, united and combined in a co-ordinated effort, incessantly lend each other mutual support. They dovetail into one another in all the daily manifestations of cerebral life, the one borrowing from the other the elements which it needs; so that from this intimate *consensus*, this co-operation of all the vital forces of the nervous elements, laid under requisition in their totality, emerges a new notion, of which we have, so far, but sketched the genesis—a notion of a whole, which is, in a manner, the synthesis of all our mental activities; that is to say, the notion of our own *personality*. Upon this subject we are now about to enter.

CHAPTER III.

GENESIS OF THE NOTION OF PERSONALITY.

THE notion of our essential personality—that *notio princeps* around which all the phenomena of our mental activity revolve—arises, as we have already hinted, from the intimate contact between the sphere of psychical activity and the intellectual sphere. It is a complex phenomenon, which undergoes development; a true physiological process which has its phases of evolution, its own mode of origin, its manifold conditions on which its life and endurance depend, and its passing moments of disturbance during which it may be eclipsed and momentarily disappear.

The elements of the vegetative and sensitive sensibility of the living organism* enter as primary factors into the genesis of the notion of our personality, and the effective participation of the elements of the *sensorium* completes and perfects it.

We have already shown, indeed, that by means of the nervous system, the elements of sensibility may be directed and drained away from the regions where they originate, and transported to a distance into the plexuses of the *sensorium*, which are the common reservoir of all the partial sensibilities of the organism.

* See p. 105, &c.

We have shown, also, that all the sensitive regions of the human organism find in this *sensorium* a symmetric point vibrating in unison with them, and that by this means our individuality in its totality, sensitive fibre by sensitive fibre, is transported to the plexuses of the *sensorium* where it is manifested.

The result is, that these plexuses enclose in their minute structure our living and feeling personality all complete, the sensitive elements which constitute it being fused into an inextricable unity. They serve as the basis of its manifestations, they unite to bring it to birth, they vivify it incessantly by their own energy, and thus, by always maintaining its vitality and sensibility, they keep it in perpetual contact with the excitations of the external world, which every instant flow in.

Through this subtle mechanism, the notion of our personality comes to life in us, being necessarily derived from a series of regular phenomena of the life of the nervous system. All the diffuse sensibilities of the organism, each in its own key, are, as we see, united in the plexuses of the *sensorium*, and thus become the primary materials for its formation.

As a natural consequence of this physiological evolution, from the very fact that the perceptive regions of the *sensorium* have given it birth, it results that it comes into direct contact with external impressions, and is inevitably associated with all the nervous excitations these develop in their train. It is constantly informed of these, is constantly *conscious* of all that passes, of the different characters and degrees of intensity of these excitations. It is impressed, it is moved, it is sorry or glad according to the various modes in which the elements of the *sensorium*, which are its natural basis, are themselves impressed by the incident stimulations.

Thus the phenomena of *conscious perception*, looked at from the physiological point of view, come within the natural limits of regularly accomplished nervous functions. There is a vital operation, a normal process, which originates and is developed by the mere fact of the co-operation of all the vital forces of the nervous system laid simultaneously under contribution. Like all the grand functions of the economy, the process on which the notion of the conscious personality depends, only lives and is maintained by the incessant concurrence of all the nervous apparatuses which take part in it; and this notion only becomes paramount and stable in itself by the continual operation of the organic mechanism by means of which it is developed.

If an interruption in the arrival of external sensitive impressions in the *sensorium* occur, special disturbances will appear, and will reveal themselves in a very characteristic manner. Thus we meet with certain patients who, when affected with anæsthesia of the lower limbs (certain forms of locomotor ataxia), say that when they are lying in bed they can no longer feel their limbs; they do not know where their legs are. They are no longer conscious of that portion of their personality which is constituted by their inferior extremities.

When the currents of blood which carry life to the cells of the *sensorium* are suspended, another order of very significant phenomena is developed. There is a sudden arrest of the working of the living machine. Everything stops at once; everything immediately remains suspended. The perceptive regions of the *sensorium*, struck, in a manner, with asphyxia, are all at once deprived of the property of feeling excitations from the surrounding medium; they remain torpid, inert, and the human personality ceases at the same time to be conscious of the things of the external world, of which it thus loses the knowledge (syncope, fainting, epileptic vertigo).

Again, if the plexuses of cells in the cortical substance, which are to a certain extent isolated as regards the arrival of blood in their tissue, as they are as regards dynamic activity, receive at a given moment more blood than is their wont, and thus assume a condition of morbid erethism, the regions of conscious personality remaining comparatively unaffected, a strange phenomenon will result, in which the individual, without having lost consciousness of external things, will be almost passively hurried away by the automatic activity of certain regions of his brain, which will urge him to utter words and to commit extravagant actions, and this in an irresistible manner, and quite without the agency of his will.

These facts lead us to the opinion that the phenomena of conscious perception, like true physiological processes, are decomposable by analysis into successive phases, and that they only develop and come to perfection through the integrity of the different media which give them birth.

From this point of view they are quite comparable to the phenomena of hæmatosis, which take place in an incessant and continuous manner only by means of the effective co-operation of a series of apparatuses of organic life, for this common end. Hæmatosis is, in its essence, a fundamental operation as important in the sphere of the phenomena of organic life as conscious perception in that of the phenomena of psychical activity. In the first case it is the arrival of the oxygen that comes to animate the blood-corpuscle and render the venous blood ruddy at the moment of its passage into the pulmonary tissue. In the second case it is the incessant and uninterrupted arrival of stimulations from the external world which animates the cerebral cell and excites its latent energies. In both cases it is the non-interruption of the arrival of the external element which is the cause of the perpetual maintenance of the function which occurs as its consequence; so much so that the notion of *conscious personality* is, in its essence, a phenomenon of vital order which exists only through the continuity and co-operation of the nervous apparatuses laid under contribution.

CHAPTER IV.

DEVELOPMENT OF THE NOTION OF PERSONALITY.

LIKE all the operations of the organism in action, the notion of our conscious personality does not all at once arrive at the degree of complete perfection which it presents in the adult. It passes through successive phases of development; it is at first rudimentary in the individual just born, and it follows by degrees, in its natural development, the successive progress of the evolution of the nervous apparatuses which are its basis.

During the first period of the life of the infant it is vague, indefinite, and as confused as the organic machinery which produces it. The plexuses of the *sensorium* are scarcely formed, cerebral biologic development waits upon that of the spinal axis, so that the automatic life then reigns alone.

It is only little by little, by means of the development of the sensorial apparatuses and those of the cerebral activity, that the infant comes to distinguish his sensations, to see, to hear, and to keep a *conscious* memory of impressions perceived. At the same time he sees himself, feels himself walk and move, has the conscious notion of his own activity; and what is more, he feels what things have pleased or displeased the sen-

DEVELOPMENT OF THE NOTION OF PERSONALITY. 239

sitive regions of his organism, and have in any way provoked the intervention of his personality.

On the other hand he touches and sees surrounding objects; he feels that all that surrounds him is not himself, that it is all external to him and his individual sensibility. Henceforth an incessant labour begins insensibly in his mind; a natural selection takes place in the mass of the acquisitions he has made, and while all the impressions radiating from the sensitive regions of his organism are fused into a homogeneous notion in his *sensorium*—the essential notion of what is himself, of his own personality—impressions from the external world, also perceived in the *sensorium*, are and remain isolated, forming a heterogeneous store, entirely apart, and henceforth classed as a contingent of external origin, independent of the former.

At this moment the young child, whose *sensorium* has, by its mere vital force, accomplished this first selection from the natural excitations which have impressed him, is (to make use of a comparison we have previously employed) in the situation of a person placed in a dark chamber, who sees his own image represented on a screen with that of external objects, and who cannot at first recognize his features nor abstract them from the objects he sees imaged on the screen. Henceforth in the mind of the young child in process of development, the phenomena of subjectivity and objectivity have an isolated existence.

This, however, is but the first step. Other operations of as great importance will soon begin—his sensibility will reveal itself externally, he will begin to speak.

240   THE BRAIN AND ITS FUNCTIONS.

This work of natural selection between internal and external impressions takes place unconsciously and in silence in the brain of the young child; his cerebral activity is not yet brought into play with all its riches; he outwardly expresses but few of the things which take place within him. It is only by degrees that he advances in the direction of mental progress. His ear at first teaches him to repeat the sounds that strike upon it, and this at first automatically, like an echo; then his mind takes its part, and his faithful memory teaches him that sounds modulated in a special manner express such and such an external object, and that accordingly the different emotional conditions of his *sensorium*, his joys and sorrows, may be outwardly expressed by significant vocal consonances. Thus, step by step, and effort by effort, he attains to the formation of a series of abstractions, and to the comprehension that if articulate sounds may be the representative signs of surrounding objects, his whole personality—his sensitive and impressionable *ego*—may be represented by a similar abstraction in a single word, by a specific sound which epitomises it, a proper name.

Thus from the earliest period of life the proper name of each individual, stamped upon the mind while it is in the act of accomplishing its first operations, becomes incorporated with its substance, and becomes for the individual and his fellows the social characteristic by means of which he passes through life. This characteristic he leaves to his successors as a hereditary patrimony, and they in their turn transmit it to their descendants with the proper attributes of genealogy.

These first acquisitions once made, the child, while conscious that he can outwardly express his emotions and desires, that he has a proper name which expresses his personality, only achieves the various degrees of his further perfectionment by a series of endeavours. At first he stammers out his first desires by means of incorrect expressions, a rudimentary attempt made up of common words. He understands the appeals that are made to him, and knows when they are addressed to his personality. In an objective excitation from without he perceives that his name is pronounced, and that he is addressed. But at the same time a very remarkable fact may be observed, which shows in a simple manner the phases through which the notion of personality passes before arriving at its period of complete *solidification* in the mind. In following these phases step by step we perceive that the child in his means of extrinsic expression, only by degrees gives up the primordial characteristics of objectivity which mark the first periods of his development.

Thus young children, about their second and third years, in the regular course of their development speak as they feel. They are accustomed to see themselves as a body which has an external form, and occupies a determined position in space. Their name itself is not as yet completely assimilated by, and incarnated in them, as the concrete expression of their entire being. They still preserve a certain degree of objectivity; in the primitive form of their language they speak of themselves in the third person, as though the matter concerned some one who was a stranger to them, manifesting their emotions or desires according to this simple formula: " Paul wishes for so and so, Paul has a pain in such or such a place."

Little by little, in the natural progress of development which is going forward, the child, living in an attentive medium, and automatically hurried along in the current of conversation, makes one step more in the way of his intellectual perfectionment.

He already knows that his personality has a proper qualification. He knows how to recognize this when it is mentioned, and turns his head and eyes when his name is pronounced; and his language, moreover, as has just been said, in a rudimentary fashion makes use of the impersonal formula. It is only little by little, and as it were by the incessant action of a continual trituration, that he can be taught that his whole personality, constituting a unity, may take another abstract form besides that of a proper name, and that its equivalent formula is represented by the words *I, me*. By a new effort of abstraction the child, who receives into his voracious mind everything that is thrust into it, unconsciously receives that conventional nutriment furnished him ready prepared, and as this is suitable, saves trouble, and is generally employed, he appropriates it, makes use of it, and by degrees employs it in current conversation. He ends by substituting the words *I* and *me* for his proper name, in the construction of the phrases he puts together according to the rules of grammar.

Having passed through this phase of mental development, which is only completed in an insensible manner, by means of a daily apprenticeship which is in force at every moment, his personality accepts the regular

DEVELOPMENT OF THE NOTION OF PERSONALITY. 243

methods of expressing itself outwardly in a methodical and regular manner, which shall be comprehended by those who surround him. It is externally clothed in a ·specific denomination, which characterizes it as a social individuality, by the proper name of the family from which it springs. It is confirmed, grows to be part and parcel of social intercourse, becomes incarnate —in a word, a precise formula which is accepted by all ; the *I* and *me* thus becoming the extrinsic grammatical manifestation of all his desires and emotions.

CHAPTER V.

FUNCTIONAL DISTURBANCE OF THE NOTION OF PERSONALITY.

IT results from the explanations we have just given that the notion of our sentient *ego*, our inner personality, far from being a simple and unique phenomenon, is merely the result of a series of organic operations, which combine and lend each other a mutual support, but which are nevertheless capable of isolated action. We have also shown that this notion of our personality, connected with the life of the organisms which sustain it, cannot maintain itself in us, always vivid and always brilliant, like a burning fire, except under the express condition that it shall be incessantly kept alive by the vital forces of the elements which concur to produce and maintain it. As it is dependent upon the oscillations of the substratum which supports it, it is liable to become languid, to rise and fall with this. We shall now endeavour to give a sketch of these different vicissitudes.

Thus it is sometimes the peripheral regions of the nervous system that are first disturbed in their functionment, and thus induce a morbid slackening in the evolution of the processes of the central regions.

Sometimes, on the contrary, it is the central regions

which are engaged, either by congestions or by sudden arrests in the course of the blood in their plexuses; phenomena which in the first case lead to exaggerations of the personality, in the second to transient obtuseness, loss of consciousness, etc.

In the first case we meet with patients (otherwise predisposed) attacked by anæsthesia of the skin, in whom sensory excitations, instead of developing in the *sensorium* the habitual reactions which result from contact with the external world, cease to react. Then we see a series of delirious conceptions of a peculiar kind occur in them; the process of personality, deprived of its elementary materials, naturally undergoes an arrest of development. Thus they think they have lost their personality, that they are changed into animals, that they have become inanimate things—a lump of clay, glass or butter—that they are no longer alive. An anæsthetic patient described by Michéa, said that his body had been changed, and that he had been transformed into a machine: "You see," he said "that I no longer have a body." Another insisted that he was dead from head to feet.* The elder Foville reports the case of an old anæsthetic soldier who said that he had been long dead. When anyone asked after his health, he would say: "How is Père Lambert, do you ask? He is dead. He was killed by a bullet. What you see is not he; it is a machine they have made to resemble him." In speaking of himself he never said *me* but *that*.

A lady affected by emotional exitement, whom I have had occasion to observe, and who was similarly anæs-

* Michéa, Annales "Médico-psychol." 1856, p. 249.

thetic, has told me she no longer felt anything surrounding her, that she was in space, that her body no longer possessed weight, and that she was on the point of flying away.

The surgeon Baudeloque, at the last period of his life, had lost consciousness of the existence of his body. If he were asked : " How is your head ? " he would answer: " My head! I have none." If he were asked to hold out his hand and have his pulse felt, he would say that he did not know where it was. He wished one day to feel his own pulse ; they placed his right hand over his left wrist, and he then asked if it were really his own hand he felt.*

When the central regions of the nervous system are engaged in their essential constitution, the most interesting disturbances may take place in connection with the processes of the notion of personality ; these disturbances being different, according as the organic conditions of the substratum differ as regards erethism or collapse of the cerebral cells, and as regards acceleration or slackening of the blood circulating in their plexuses.

Thus in the congestive period of general paralysis, when the elements of the *sensorium*, suffering from the most intense circulatory super-activity, receive nutrient materials in excess, they are by this means impelled, like all the other histological elements of the economy, to develop their peculiar vitality in an exaggerated manner. They then assume a condition of exaltation, and soon develop a species of continuous erethism,

* Quoted from Michéa, *loc. cit.*, and " Bibliothèque Médicale," 1809, vol. xi.

DISTURBANCE OF THE NOTION OF PERSONALITY. 247

while the physiological function they consequently accomplish by no means increases in an equal proportion. Thus it is that in this congestive period the normal process of the evolution of personality is exaggerated in so characteristic a manner. At this period, indeed, the personality of the individual is raised several degrees above its normal pitch. It extends, enlarges,* swells out, with the morphological elements upon which it lives, and the patient, hurried into that fatal cycle, feels himself richer, greater, stronger than he was before. He speaks of himself, his physical health, which is splendid, of the riches he has accumulated, of his social importance, which is immense—he has become a king, an emperor, a pope, etc.

Under contrary conditions, when the plexuses of the *sensorium* no longer receive a sufficient quantity of blood, as regards their assimilative properties, inverse phenomena are produced.

The elements of the *sensorium* are affected with a species of general torpor which causes their vital energies to sink below their normal pitch, and they accordingly exhibit that general condition of diffused languishing of the mental forces, in which the processes of personality are only manifested in a dull, vague, and diffuse manner. The patients, then a prey to certain forms of melancholy with stupor, present a more or less complete passivity, an apathy and profound indifference for all that comes in contact with them; and usually this torpid condition is only the return effect of a sort of anæsthesia of the central regions

---

* I have seen a patient, in the congestive stage of general paralysis, who assured me every morning that he had grown a foot higher.

which goes hand in hand with that of the peripheral regions.*

There is still another series of morbid phenomena in which the notion of personality, and consciousness of the external world may be suddenly suspended by the effect of a momentary arrest of the circulation in the plexuses of the *sensorium*.

We now know, thanks to the labours of modern physiology, that intra-cephalic circulatory disturbances are frequent in epileptics, and that at the moment of the attack the loss of consciousness is produced by a spasm of the vessels, which interferes with the course of the blood through the cerebral substance. It sometimes happens that these circulatory disturbances, far from taking place throughout all the extent of the regions of the *sensorium*, as at the moment of the great epileptic attacks with complete loss of consciousness, exercise their influence only within limited regions of the cerebral substance. There are then local arrests of circulation in certain cell-territories, which are for the moment in a state of collapse—true partial ischæmias — while in others the cerebral activity continues its function in an independent manner. We see individuals, as if in a state of somnambulism, act unconsciously, commit extravagant actions, even crimes, without having any conscious idea of the things of the external world ; and at the end of several hours, or even of several days, emerge from this condition of

---

* In a patient affected with melancholia with prolonged stupor ending in death, I succeeded in discovering a most characteristic condition of anæmia of the cerebral substance, which was as it were washed clean and deprived of sanguine materials.

DISTURBANCE OF THE NOTION OF PERSONALITY. 249

partial stupor of their *sensorium*, quite astonished and stupified by the words they have pronounced and the deeds they have done during this period of interregnum of their conscious personality (unconscious alienations).*

Finally, we may remember that the notion of our personality, which in its constitution and its very existence is under the jurisdiction of the organic machinery in the midst of which it lives, is regularly eclipsed every twelve hours, when the cerebral cells relapse into the condition of sleep.

The cerebral cell, in fact, like the peripheral cell (sensorial cells of the retina), becomes fatigued at the end of a certain period of activity; its sensibility becomes more or less rapidly dulled. It is fatigued, and perforce falls into a state of collapse, which is nothing but physiological sleep. At this period it ceases to attract blood to it, the circulation slackens, and in proportion as the period of sleep becomes better and better marked, and *loss of consciousness* of surrounding circumstances occurs, the *notion of our personality* at the same time grows dull, and finally becomes extinct, and this in a more or less complete manner, according to the temperament and habits of each person.

* See the cases of transitory mania reported in my work on the "Cerebral Reflex Actions," p. 137.

BOOK II.

PHASE OF PROPAGATION OF THE PROCESSES OF CEREBRAL ACTIVITY.

CHAPTER I.

DISSEMINATION OF SENSORIAL IMPRESSIONS IN THE PLEXUSES OF THE PSYCHO-INTELLECTUAL SPHERE. GENESIS OF IDEAS.

WE have already seen that sensorial impressions, once received into the different regions of the cortical periphery, become dispersed in the plexuses of the *sensorium*, which constitutes for them a vast field of projection, and that, pursuing their course from this point onwards, they enter into particular relations, some with the sphere of psychical, others with that of purely intellectual, activity. In these cerebral regions they find the last stage of their long migrations through the organism. There they are concentrated and transformed, and, under new forms, having become *intellectualized* excitations of the psycho-intellectual sphere, they constitute the fundamental elements of all the phenomena of cerebral life.

There, in fact, these same sensorial excitations, incarnated in the living cell, become perpetuated as persistent excitations; to become, as it were, durable memorials of the first impression that gave birth to them. There they repose, in those infinite labyrinths of the psycho-intellectual sphere where they live, always alert, always brilliant, like faithfully-kept archives of the past of our intellect and emotions. There they form that common fund of ancient memories, accumulated from our earliest years, which gives birth to those *fundamental-ideas* which we always carry within us, and which are but radiations from the external world, that have previously been impressed upon us. They have lived with us for long years, and have assumed in a manner an independent existence, like foreign grafts implanted in our substance. The ideas and emotions which are nearest to us are, then, only direct reflexions and prolonged repercussions of the external world that have impressed us during our course through life; and this subtle operation, which commences with the earliest phases of our existence, is perpetuated, and perpetuates itself incessantly, by an incessant participation of the brain's own activity.

Each sensorial impression that affects us leaves a record, a specific memory; and it is this posthumous memory of the absent object that continues to vibrate, that perpetuates, vivifies, reinforces itself by means of excitations of the same pitch, which communicate to it a new freshness when it begins to grow feeble. The origin and permanence of our ideas, as of our emotions, depend upon this daily maintenance of persistent impressions.

If, indeed, we inquire profoundly into the genealogy of each of these in particular—if we submit each to a series of elementary analyses, decomposing it into its primary elements, we shall always find as the ultimate result, at the bottom of the crucible, that our ideas, like all our emotions, are reducible to a sensorial impression, as the fundamental condition of their occurrence. This sensorial impression is at the bottom of all our ideas, all our conceptions, though it may at first conceal itself in the form of a binary, ternary, quaternary compound; and, on our methodically pursuing the inquiry, it is easily recognizable—just as a simple substance in organic chemistry may always be summoned to appear, if we sit down with the resolution to disengage it from all the artificial combinations which hold it imprisoned.*

---

* The ideas of time and space, which philosophers have so long considered irreducible, are, however, decomposable by analysis into simple elements which connect them with the regular processes of cerebral activity.

Thus the notion we acquire of space is directly derived from that of muscular activity. It is by the notion of the amount of the effort made to change our position that we acquire the notion of the road passed over, and of its length. It is by steps that blind men judge of the distance from one place to another, and thus acquire the notion of space. It is thus that we successively appreciate mentally the space occupied by a yard, a mile, several miles, etc.; and we finally arrive at the conception of the immensity of the interplanetary spaces. It is therefore the data we have acquired from objective nature that preside over the construction of our notion of space.

The same holds good for the notion of time. It is a very complex process into which many factors enter, and above all the registry of daily facts. Thus as regards the appreciation of the hours of the day, we refer to the intensity of the brightness of day, and the repetition of habitual incidents, and to many conditions of the medium surrounding us which periodically occur at a given moment. We recognize months and years, by the facts of our memory and notes which register what we have done. This is so real, that when the elements of the *sensorium* are disturbed, when the memory becomes impaired as regards the retention of recent facts, the notion of time disappears. A great

GENESIS OF IDEAS.

All our ideas and emotions originate then, physiologically, in an external phenomenon which is incarnated in us, and perpetuates itself as a remembrance; and it is thus that our ideas, like our remembrances, live in the life of the organic substratum that supports them, and with it undergo all the oscillations that may affect it.

Thus by means of the calling into activity of the nerve-cell with all its intrinsic and extrinsic attributes, the sensorial impression imprinted upon it becomes an *idea*, that is to say, a remembrance of the absent object. It is propagated to a distance by means of anastomotic plexuses, and is thus transformed, by cell after cell, into a progressive and radiating impression.

Thus, by means of these connections, our ideas are associated, grouping themselves methodically into contemporary reminiscences, appealing one to another, when the first link of the chain has been struck; presenting themselves again in an irregular and disconnected manner when, abandoning the direction of our mind, we let it run wild, as it is termed ; when we give audience to our thoughts, that is to say, when we leave the automatic activities of our cerebral cells to exercise themselves according to their natural propensities and appeal to one another according to their natural affinities.

It is by means of this organic mechanism that movement and life are incessantly spread through the plexuses of the cerebral cortex ; that excitations of all kinds spring up in their minute structure on the arrival of

number of lunatics who have been for several years in asylums, take no note of the time that passes, and make considerable mistakes respecting this ; they say they have been shut up for five or six years, when the time of their entry into the establishment dates back as much as fifteen or twenty years.

external impressions; that the materials of the past become associated with recent ideas and impressions, and that, in a word, those marvellous phenomena, so instantaneous and so varied, presented by the activity of the brain, are developed in presence of the conscious personality, which assists, as a spectator, at their evolution, without being able to direct the movement which is accomplished, and, strange to say, with the idea that it is regulating them.

We generally imagine that we ordain the direction of our ideas into any desired channel, and that we can govern their evocation. We do not usually perceive that, while we imagine we are leading our ideas in one direction, we are unconsciously obeying the second phase of a movement of which the first has already taken place.

I imagine that I think of an object by a spontaneous effort of my mind; it is an illusion—it is because the cell-territory where that object resides has been previously set vibrating in my brain. I obey when I think I am commanding, merely turning in a direction towards which I am unconsciously drawn. A phenomenon quite analogous to the conjuring trick of forcing a card takes place in this instance; the conjuror forcing us unconsciously to take a card, while letting us imagine we have a liberty of choice.

Sensorial excitations, once they are disseminated in the plexuses of the cortical substance, continue, as we have already several times said, the movement commenced by their contact with the external world. The process in evolution pursues its course, and then they are distributed—some to the sphere of psychic activity,

GENESIS OF IDEAS.

others to that of intellectual activity proper. We shall now pursue the study of them into these two regions.*

* Peripheral impressions do not all arrive at the *sensorium* with equal rapidity even in the same individual. In the sensitive nerves the rapidity of transmission has been variously estimated. It oscillates according to different authors, between 24 and 26 *metres* a second. It is modified by several influences, cold for instance and the electro-tonic condition. It is probable that it is not uniform, and that it diminishes in the ratio of the distance of its origin. (Hermann, "Physiology," p. 319.)

Mach has endeavoured comparatively to determine the minimum time for the conversion, within the brain, of an impression into a motor excitation. For visual impressions the rapidity of transmission is 0·0472 ; for tactile impressions 0·029; for auditory impressions 0·016. That is to say, of all impressions auditory are most rapidly perceived. ("Annales Médico-psychologiques," 1869, vol. ii. 6, 441.)

On the other hand, astronomers have long designated under the name "individual coefficient," that allowance which must be made, in correcting formulæ, for the unequal rapidity with which different observers perceive the occurrence of the same celestial phenomenon. We know, in fact, that when several persons are charged with the notation of the exact time of the meridian passage of a star, there is never perfect synchronism between all their observations. The transmission to the *sensorium* of the luminous impression, and its conversion into a reflex motor excitation, takes place with unequal rapidity in different persons. This leads us to the conclusion that, as regards intellectual operations, there is a physiological habit proper to each individual ; that there are persons slow to see, slow to comprehend and to react, just as there are, as regards the phenomena of somatic progression, persons slow to move, and lazy in walking.

CHAPTER II.

EVOLUTION AND TRANSFORMATION OF SENSORIAL IMPRESSIONS.

Evolution of Sensitive Impressions.—Sensitive impressions in general comprehend not merely impressions of touch, contact, and pressure of bodies, but also those which give us the idea of temperature, and that of the activity of our muscles. They are designed, either isolatedly or simultaneously, to play a principal part in the phenomena of cerebral activity proper; forming, as has been explained, an enormous contingent of connate excitations which are distributed in the domain of psychical activity proper as well as that of intellectual activity.*

Radiating from the central regions of the optic thalami which represent the very centre of the brain, they do not as yet appear to have a very clearly defined localization, as regards their ultimate distribution. Indeed, the fibres that radiate from the median centre appear as though they must distribute them equally throughout the different zones of the cerebral cortex.

* The part played by sensitive impressions in the phenomena of cerebral activity is so important, as regards the physiological stimulation they develop, that when, in consequence of amputation of the limbs, they have long ceased to stimulate the brain, the hemisphere that has ceased to receive them undergoes a correlative atrophy, in consequence of the cessation of their influx.

The contingent of sensitive elements specially reserved for distribution in the field of psychical activity, as we have defined it, is represented by all those agglomerations of sensitive excitations which, drawn from all the sensitive points of the organism, are conducted towards the central regions by the centripetal channels.

These agglomerated sensitive elements, incessantly vibrating with one accord, incessantly active, become in the *sensorium* the elements constituting our inner personality, our sentient unity. This is the special part played by sentient impressions as regards psychical activity proper; and we see what an important part it is, they being the keystones of the whole edifice of our mental activity, since they produce by their synthesis the notion of a living individuality in exercise.

Genesis of the Notion of Happiness and Unhappiness.— Sensitive impressions are again reverberated in the *sensorium* in a very peculiar manner, exciting in it conditions which depend on them alone.

Thus from that pre-established consensus between the peripheral and central regions of the nervous system, on which we have so strongly insisted, this very remarkable consequence results: that the special condition of the sensitive nerves (when affected by impressions which gratify their natural sensibility) is reflected upon the *sensorium*, and there develops a species of concord, by means of which it enters into unison with them.

When a warm atmosphere refreshes our skin with gentle perspiration, when comfortable repose revives our strength and restores to our fatigued muscles and aching joints their pristine flexibility and elasticity, we say that

S

we are in a special condition of comfort—that this has given us *pleasure*.

This word *pleasure* characterizes a special state of our *sensorium*, a peculiar pitch of the sensibility, which is desired by every one, and which thus becomes a specific mode of existence of the *sensorium*, which fixes and perpetuates itself in us as a memory and a hope. It is a kind of specific sentiment, a species of standard sentiment with which we compare the greater number of the impressions that come to be reflected in us ; so that, by extension, the notion of the pleasure of our gratified sensitive nerves insensibly becomes subjective, to be transformed into the notion of *happiness*. It results from this mental evolution that when any act whatever of the human activity is judged of by us, we say that it is *good*, because it has produced in the sphere of our moral sensibility an impression equivalent to that produced in the domain of physical sensibility by a sensorial impression which has given us *pleasure*. And, inversely, whatever wounds or offends our physical sensibility—whatever gives us *pain*—places our *sensorium* in very different conditions from the foregoing, and thus becomes the subjective notion of *unhappiness*, to which we refer all the miseries of our moral sensibility.

In the domain of intellectual activity proper, sensitive impressions also come to be of the utmost importance.

United with the correlated impressions that emanate from the minute structure of our muscles when in action, they make part and parcel of a number of complex notions, by which the understanding profits, and which are incessantly laid under contribution without our having any clear consciousness of the fact.

It is chiefly tactile impressions that form the special contingent destined to provoke the reactions of the intellectual sphere.

Radiated from the extremities of the peripheral plexuses, gifted with a special organization (sensitive papillæ, tactile corpuscles of Pacini), these impressions furnish the intellect with a number of notions, not very numerous, it is true, but very precise, respecting the different qualities of bodies in contact with them. It is by means of them that we form our judgments respecting the dimensions and surface-condition of external bodies, and respecting their motion, temperature, and degree of dryness or moisture. It is by means of them and their fellows of muscular sensibility that we are informed of the expenditure of nerve-power necessary to gauge the weight of heavy bodies, to lift them, and indirectly acquire a precise notion of their volume and solidity.

This special contingent of sensitive elements, by means of which the notion of human personality is developed and maintained, and by means of which also we are constantly in contact with the things of the external world—this contingent, I say, is still destined to vibrate in harmony with all the mental faculties, and to give specific bent to the character of the individual, as well as to the creations of his mind. We may say, then, that a greater or less degree of perfectionment, and a greater or less degree of sensitive power in the sensitive regions, find their counterpart in the central regions, and that the greater the degree of physical, the greater will be the degree of moral sensibility.

We all know how fine, delicate, and sensitive is

the skin of women in general, and particularly of those who live in idleness and do no manual work—how their sensitive nervous plexuses are in a manner exposed naked to exciting agencies of all sorts, and how, from this very fact, this tactile sensibility, incessantly awake, and incessantly in vibration, keeps their mind continually informed of a thousand sensations that escape us men, and of tactile subtleties of which we have no notion. Thus in idle women of society, and men with a fine skin, mental aptitudes are developed and maintained in the direct ratio of the perfectionment and delicacy of sensibility of the skin. The perfection of touch becomes in a manner a second sight, which enables the mind to feel and *see* fine details which escape the generality of men, and constitutes a quality of the first order, *moral tact*, that touch of the soul (toucher de l'âme), as it has been called, which is the characteristic of organizations with a delicate and impressionable skin, whose *sensorium*, like a tense cord, is always ready to vibrate at the contact of the slightest impressions.

Inversely, compare the thick skin of the man of toil, accustomed to handle coarse tools and lift heavy burdens, and in whom the sensitive plexuses are removed from the bodies they touch by a thick layer of epithelial callosities, and see if, after an examination of his intellectual and moral sensibility, you are understood when you endeavour to evoke in him some sparks of those delicacies of sentiment that so clearly characterize the mental condition of individuals with a fine skin. On this point experience has long ago pronounced judgment, and we all know that we must speak to every one in the language he can comprehend, and that to endeavour to awaken in the mind of a man of coarse skin a notion of the delicacies of a refined sentiment is to speak to a deaf man of the deliciousness of harmony and to a blind man of the beauties of colours.*

Evolution of Optic Impressions.—The luminous vibrations, directly transformed into nervous vibrations by the peculiar action of the retina, are all at first concentrated in the grey centres of the optic thalamus devoted to them, and radiated thence, chiefly into the antero-lateral regions of the cerebral cortex. They arrive in the *sensorium*, as we have already described, with different degrees of rapidity in different individuals,† and from the time when they come in the morning to illuminate the nervous plexuses of the *sensorium* they are continuous, and by their incessant stimulation during the period of waking maintain the activity of the cerebral cells in continued erethism.

The luminous undulations which thus radiate through the brain are not homogeneous as regards their ex-

* To the facts we have already cited respecting the pathogenic influence exercised by certain anæsthesias upon the genealogy of certain forms of delirium, we should add as a complement the following observations reported by Dr. Auzouy, which clearly show what a curious influence sensitive impressions may have upon psycho-intellectual phenomena in general. The case was that of a young man, clever and rational, who suddenly became undisciplined and rebellious to the utmost extent, and gave himself up to the worst tendencies, even to the compromising of the peace and honour of his family. Examination showed that he was completely anæsthetic. During his stay in the asylum he successively experienced several intermittent phases of anæsthesia, of which the appearance manifestly coincided with the return of his worst instincts. When sensibility reappeared in the skin, moral dispositions contrary to the preceeding were observed to return in him, together with a very clear consciousness of his situation. (Auzouy, "'Annales Médico-psychol.," 1859, p. 535) *Des troubles fonctionnels de la peau et de l'action de l'électricité chez les aliénés.*

† See p. 255 (note).

trinsic characters, and do not equally affect the different regions of the cortex in which they are distributed. Thus, not only do they transmit to the *sensorium* perceptions of the different gradations of intensity of light, but furnish as well the most specific notions of the colour of surrounding objects. There are thus, in fact, two different modes in which the elements of the *sensorium* may be affected; and in most men one or other of these modes usually predominates. We meet with certain organizations which from this point of view are very unequally endowed. Every one knows that if all persons with the gift of sight have the faculty of being impressed by light, all have not the faculty of perceiving colours in an equal degree, and that there are persons who suffer from a peculiar form of blindness which makes certain hues virtually non-existent for them.* We all know that certain painters, who are gifted in the highest degree with that natural aptitude for perceiving in a complete manner the different gradations of the colour of objects, can give to their works a quite unique intensity of colour, a richness of tone which they draw from their own personality, and which their less gifted rivals can neither comprehend nor imitate.

Optic impressions, as well as sensitive, are divided into two contingents which are separately distributed,

* Mr. Black saw a man of fifty years of age in Glasgow who had lost his sight when two months old, and yet learnt by degrees to distinguish colours so clearly that he could exercise his profession as a dyer, without any help, for more than forty years. He could not only perfectly appreciate colours and shades, but had learnt by practice to give the stuff a lighter or darker tint without making any mistake. ("Annales Médico-psychol.," 1848, p. 414.)

See also the memoir of Earle on the incapacity for distinguishing colours. "Annales Médico-psychol.," 1846, p. 217.)

EVOLUTION OF SENSORIAL IMPRESSIONS. 263

either in the sphere of psychical or the sphere of intellectual activity.

1. *Genesis of the Notion of Beauty and Ugliness.*

The particular contingent of optic impressions destined to be distributed in the sphere of psychical activity appears to be the origin of that faculty by which we pronounce as to the *beauty* or *ugliness* of the thing that impresses us, and in this it resembles those sensitive impressions that furnish us with the notion of *happiness*, by means of a regularly accomplished physiological process. These optic impressions are similarly the fundamental impressions that engender in us the notion of the *beautiful*.

These optic impressions, indeed, originating as they do, like those of general sensibility, in the peripheral regions, do not ascend into the *sensorium* in the condition of atonic, indifferent, slightly-stimulating impressions. They carry with them the special condition into which the peripheral plexuses have been thrown at the moment of their genesis, and the simultaneous notions of concomitant pleasure or pain. When an agreeable spectacle presents itself to our eyes, our retinas, being impressionable nervous plexuses, are more or less directly gratified as regards their natural sensibility, just as when an agreeable sensation affects our sensitive nerves; and this special satisfaction is transmitted to the *sensorium*, thereby producing in it also a special vital condition, a new state which we express under the denomination of a *sensation of beauty.**

* Thus there are intrinsic satisfactions for the eyes as well as the ears. It is with infinite pleasure that we all salute the light on emerging from obscurity; that our eyes are pleased to receive the primitive rays of the spec-

The subjective notion that we have of the beauty of things is thus, in the primitive man, who knows nothing of either the subtleties of art, or the casuistries of the different schools, or the code of amateurs, fundamentally connected with the memory of an agreeable impression, a purely visual satisfaction felt by the retina when agreeably affected.

Children love all that is brilliant and that glitters in the sun; the inhabitants of northern countries and certain savage tribes, are attracted by the sight of objects of a vivid colour, and tints which violently affect the sight. These are the rudimentary forms of the idea of the beautiful, which is really derived from a primitive physical impression. It is only by degrees, by means of the participation of the intellect, the culture of the judgment, and comparison, that this first notion comes to perfection in us, and becomes a rational well-digested appreciation, though having its origin in a physical impression which is at first addressed to our optic sensibility.

Conversely we can comprehend that those things which produce on the retina a painful impression, which are unpleasant to see, are also those which produce a painful impression on the *sensorium*, and which bring with them a notion the reverse of the former, that is to say that of *ugliness*.

2. Optic impressions, when carried up to the *sensorium*, not only excite in it special conditions by means of which the notion of beauty or ugliness is naturally

trum, that they rejoice in the magnificent stained-glass of our old cathedrals when the sun shines through them, in the folds of rich satin stuffs, the multicoloured reflexes of brilliant flowers, fireworks or coloured flames.

developed in us, but they are further gifted with a more intense penetrative power, and while taking upon them a thousand forms they touch and set vibrating all the chords of our emotivity.

Thus the sight of a landscape in full sunshine, enamelled with flowers of a thousand hues, and covered with green meadows with distant horizons, develops in us sentiments of satisfaction which gratify our sensibility and cause it to expand; while a gloomy place, shut in by high walls, and without verdure, saddens the *sensorium*, and develops in us a very legitimate sentiment of repulsion, in which all share. Thus these sentiments of attraction and repulsion are directly imposed upon us in consequence of the perceived impression, without the intervention of memory or of old reminiscences.

By reason of those mysterious affinities which unite the present with the past, as regards our ideas and emotions, a simple appearance, a simple optic impression, is capable of reviving old memories, and according to circumstances, of setting in vibration all the different emotional chords that it touches within us.

Thus the sight of an external symbol, a banner, a standard, a flag, is capable of suddenly exciting in those who behold and salute it, the most diverse sentiments, from the fact that its appearance awakes in them a series of individual reminiscences. It is by the sight of the external pomp that surrounds them, the display of gold and silver embroideries, of brilliant uniforms, that the possessors of authority at all times and in all places have sought to inspire respect in the crowds before which they have passed. It is by securing the passive admiration of the eyes of their dazzled contemporaries that they have always maintained their prestige. It is for the gratification of the eye that human beings over the whole surface of the globe seek, according to their means, to ornament their persons and appear to the utmost advantage externally.

It is by the lust of the eye that we are all, small or great, young or old, rustics or citizens, captivated and allured; for it is always our eyes that are first charmed by the contemplation of physical beauty; and the most powerful of sentiments, love, destined to set the heart of man beating, has, as a general rule, its sole origin in the seduction of the sight, the pleasure of the eyes, which ardently desire the object which has charmed them, and excited the spontaneous awaking of all latent delights.

It is, moreover, by means of those mysterious links which associate optic impressions with our sentiments, that our former emotions, our secret affections are awakened and maintained by the sight of certain keepsakes. Every one knows what a sweet consolation for the absent are the features of a beloved person reproduced by painting; how certain institutions, certain public or private ceremonies recurring in a periodic manner, certain anniversaries, are similarly calculated to revive in us former emotions, and again bring us into the presence of the persons and circumstances that have first inspired them, recalling the periods at which our emotions have been set in movement.

3. Again, in the sphere of purely intellectual phenomena, optic impressions play a very important part which deserves attention.

Thus, either alone or associated with their excito-motor fellows, which regulate without our knowledge the different movements of accommodation of the eye, they permit us to judge of the distance, the dimensions, and the forms of different surrounding objects. Thus, as when we have to do with the impressions of sensibility proper, former impressions are associated with recent, to form the elements of comparison. When we say that a body is at such or such a distance from us, there is a reflex action of the intelligence which, from our knowledge of the object, and the manner in which it is illuminated, associates a series of notions previously acquired with a recent impression. When, as regards a body that moves transversely before us, we judge of the direction of this movement, it is still the evocation of an impression formerly received that comes to be annexed to a recent impression.

Thus by degrees a crowd of complex notions is created in the mind by the arrival of optic impressions, and their preservation in the state of persistent memories. The sense of sight consequently becomes one of the most fertile sources from which all our cerebral activity is incessantly fed. It is optic impressions again that with their acoustic fellows are called on to play such an important part in the artificial culture of the mind, both in the mental interpretation of graphic signs in the action of writing from dictation, and in the regular tracing of such characters in the action of writing spontaneously. They are also the introducers of the thoughts of others into our minds, when, with our eyes fixed on the written characters, we attach to each of these characters correlative ideas and co-ordinated emotions. They thus animate these silent characters, giving them life and fixing them in us as materials designed to excite in the mind new associations of ideas, and the most varied impressions.

They are therefore, in fact, the most powerful agents that stimulate the culture of the psycho-intellectual sphere, and fertilize its activity. They permit us at once to receive impressions from the thoughts of others, by means of written words, transmitted to a distance, and reciprocally to manifest our emotions and ideas in a manuscript form, which thus becomes the manifest expression of the different states that they pass through.

4. The important part that optic impressions play in the functionment of mental activity leads to the conclusion that when they are wanting there will be a certain disturbance of the general equilibrium, which will have as its consequence special disturbances of cerebral functionment.

Up to the present time, the mental condition of the blind has not been studied in a sufficiently precise manner to permit of our clearly appreciating the modifications which occur in the character or fashion of their ideas, under the influence of the arrest of development of their optic impressions. Nevertheless, we may say with Dumont, who has already occupied himself with this question, that the influence that optic impressions exercise upon the play of the cerebral functions is most important, and that a certain number of individuals, whom he had an opportunity of observing, presented, from a psychical point of view, changes of temper and symptoms of melancholy, all the more marked because the patients were incapable of discerning day from night.*

As regards such phenomena, Bouisson has observed a most remarkable case.† The patient was a young man who had become insane in consequence of a double cataract, with incoherence of ideas, complete failure of spontaneity. Bouisson, from the antecedents of the patient, hit on the happy idea of performing an operation. It was simultaneously performed in both eyes, by couching, and a few days afterwards, when optic impressions reappeared to stimulate regularly the *sensorium* of the patient, and vision was restored to him, he began to utter a few sensible words, his mental state became progressively better, and, at the end of a few weeks, he left the hospital capable of attending to his own wants.

Baillarger has also reported analogous facts. Thus, he cites from Whytt the case of a patient who, if his eyes were closed by another person, even without sleeping, fell into a great disorder of mind. It seemed to him that he was transported through the air, and that his limbs were falling off.

In a patient of twenty-seven, whom he observed himself, he noticed that as soon as she shut her eyes, she saw animals, fields, and houses. " I several times closed her lids myself," he says, "and immediately she mentioned to me a number of objects that appeared to her."‡

* According to Dumont, among 120 blind persons, excluding those who are affected with appreciable brain lesions, there are thirty-seven with intellectual disorders varying from hypochondria to mania, hallucination and dementia. (Influence of blindness on the intellectual functions.) "Moniteur des Hôpitaux," 1857, pp. 245 and 265.

† "Bulletin de l'Académie de Médecine," 8th Oct., 1860.

‡ Baillarger, "Annales Médico-psychol.," 1845, pp. 22, 23. (On the influence of the state intermediate between sleep and waking.)

Evolution of Acoustic Impressions.—Acoustic impressions, like optic impressions, play a most important part in the sum-total of the manifestations of mental activity. Like them, they are incessant during the whole diurnal period, and by their uninterrupted stimulation maintain cerebral functionment in a perpetual condition of erethism. They are, for us, the natural vehicles of the notion of sound and harmony, while, at the same time, they are the generating elements of articulate language. Through them the ears are charmed, the understanding perceives and interprets, according to conventional methods, articulate vocal sounds, and the human personality thrown into emotion vibrates externally, and expresses itself in regularly co-ordinated vocal sounds.

They are collected at the periphery of the acoustic sensorial plexuses, and, like their fellows, are condensed in special ganglia of the grey substance of the posterior regions of the optic thalamus, and thence radiated, principally into the posterior regions of the cortical substance, which, in the human species, present such a characteristic development. According to Wundt, they are the impressions most rapidly transmitted to the perceptive centre.

Like their fellows, they have a double range; they enter into relation successively with the psychic sphere and the intellectual sphere proper, and in these two regions of nervous activity they excite specific reactions of the same nature as their fellows do.

1. When dispersed in the plexuses of the *sensorium* they at first develop there the same reactions of pleasure and pain that we have seen succeed each other in

consequence of the arrival of sensitive and optic impressions, according to the same physiological processes. The variable condition of impressionability of the peripheral regions is always transmitted into the central regions, and there excites concordant emotional states. When the ears are charmed, the *sensorium* is similarly delighted, and inversely when the ears are impressed with a certain rhythm and with certain modulations into flat or sharp keys, the same states are impressed upon the sensorium.

Thus it is that grave musical sounds, repeated very slowly and in a chanting manner—musical phrases in flat keys, and *andante*—dispose the *sensorium* to reminiscence, and produce in us a special condition which constitutes sorrow; and that, inversely, loud music, consisting of rapid notes, and *allegro* in *tempo*, or airs in $\frac{2}{4}$-time and tricked out with sharps, awakes emotions of an entirely different nature, predisposing the heart to gaiety and mirth, and inviting us to dance spontaneously and move our limbs to its cadence.

Between these two limits of profound sorrow and expansive joy, between which acoustic impressions cause our natural sensibility to oscillate, there is a whole series of intermediate notes which may be successively set in vibration.

Music, indeed, with its infinite number of tones, is capable of impressing us in various manners, and developing sensitive conditions very distinctly graduated. It is, like spoken language, of which it is but an amplification, designed to form a sort of synthetic language, and to join the train of the cardinal sentiments which are capable of causing the plexuses of the human *sensorium* to vibrate. Thus musical sounds now express tender sentiments, flowing forth in sweet harmonious notes, and in slow time; while in other circumstances, with that richness of expression the great masters have given to their works, we see a melodious phrase augmented by graduated accompaniments become infinitely complicated, and with the aid of powerful orchestration symbolise the most complex sentiments, not merely of man considered as a sentient unit, but even of man considered as a social unit. Thus it is that the great masters have succeeded in expressing in music the different shades of human sensibility, just as the masters of painting have done with their palette,* and in indelibly imprinting their inmost thoughts, and the sentiments with which they were animated, upon the *sensorium* of those who comprehend them.

Acoustic excitations, associated with all the special emotions of the period at which they are implanted in the *sensorium*, thus perpetuate themselves in the form of memories and as a persistent echo of the past. They are thus capable of reviving, with the qualities with which they were previously gifted. Every one knows, indeed, that a musical phrase is sufficient to recall the circumstances in which we heard it for the first time; that that instantaneous recollection of certain airs heard during childhood, which is often so vivid, is capable of awakening in us the memory of the places and circumstances in which they were first heard; and that national airs, among peoples who have imbibed

* Thus Meyerbeer has succeeded in giving a musical expression to the enthusiasms of politics and the fanaticism of religious strife, in his grandiose scores of *Les Huguenots* and *Le Prophète*.

the national sentiment in a precise formula, become very dear to those who hear them when far from their country, and are like a perfume from their distant home.

2. Besides this special category of acoustic impressions which directly address the *sensorium*, there is another contingent destined to play a most important part in the phenomena of cerebral life—that which directly serves for the manifestations of verbal expression.

In the first phases of the development of the young child, it is indeed acoustic impressions that first awaken his mind, and lead him to reproduce the sounds that strike his ears. They are stored up in his *sensorium* as persistent memories, represent the absent objects that have been named verbally in his presence, and when reproduced by a reflex action of his brain, become the natural excitants of the different phonetic expressions by the aid of which he designates the same objects, as well as the different conditions affecting his sensibility. It is by means of this series of acts that human speech, the natural daughter of auditory excitations, becomes developed in us, expresses itself outwardly, and manifests through precise and appropriate sounds the emotions of the sentient personality which is in action.

It amplifies and develops little by little, and becomes in course of time a true vital force, capable of acting at a distance like a charged electric machine, and of discharging upon the *sensorium* of another person, and modifying by its seductive influence his sensibility as well as his intelligence. By virtue of the energy with which it is projected, and the heat with which it is expressed, it is capable of provoking different emotions at a distance from the spot where it was engendered, and of exciting sympathetic and persuasive effluences which induce a tacit acquiescence on the part of whoever perceives it. It thus creates a sort of automatic consonance between the orator and those who hear him, and becomes the bond of union which links us to our fellows. It is always due to it that men speaking the same language have among them common points of contact, by which their *sensoria*, the sensitive regions of their whole personality, converse, touch each other, and vibrate in unison.

3. The special contingent of acoustic excitations which reverberates in the purely intellectual regions, becomes the origin of a series of appropriate judgments which we form respecting the timbre and intensity of sounds emanating from the different sonorous bodies around us.

Thus we judge of the specific pitch of a given sound, by dint of a phenomenon of the memory, by juxtaposing in our mind the reminiscence of a past sound of the same nature as the sound that now strikes our ear.

We judge of the intensity of a sound-producing agency by the manner in which it impresses our auditory nerves, of which the sensibility is called into play ; and perhaps the notion of muscular activity—the work accomplished by the tensor muscles of the tympanum—may play a certain part in this operation.

It is, further, by a reflex effect of the mind and the memory that we arrive at a judgment respecting the distance of a sounding body. We know that when a known sound gradually decreases in intensity, it is because the sonorous body is receding, and when, on the contrary it gradually increases, it is because the sonorous body is approaching. These two acquired notions afford materials for our judgment in a given case.*

Evolution of Olfactory Impressions.—Olfactory impressions, collected from the peripheral plexuses of the corresponding nerves, are directly transmitted, as we have already explained, to a special department of the optic thalamus, the anterior centre. We have already insisted upon the comparatively large volume of this sensorial ganglion in those vertebrates that present a great development of the olfactory nerves; upon the multiple connections it effects with the grey substance of the *septum lucidum* and mamillary tubercles; and, finally, upon the indirect relations which unite it to the regions of the sphenoidal lobe, and in particular to those of the grey substance of the hippocampus.†

The olfactory nerves transmit to the *sensorium* the specific and unanalysable notion of odours. They communicate to it at the same time a special coefficient

---

* When these relations are interrupted, the conscious personality easily accepts the change and allows itself to be hurried into strange illusions. It is by muffling the sounds that he produces, in the act of production, that a ventriloquist makes his audience believe that the sounds so produced come from a distance. It is by means of the same mechanism that phantasmagoric illusions in the domain of visual impressions make us think that an image which grows larger and larger on a flat surface is approaching us.

† The multiplicity of the paths traversed by the olfactory impressions in passing through the brain, the irregularities and individual varieties of each of the stages through which they are propagated, must exercise an influence upon their central mode of elaboration. It is perhaps only in these quite special conditions of irregularity in the transmission of olfactory impressions to the *sensorium*, that we must look for the secret of those individual varieties which we so frequently observe among individuals questioned respecting their appreciation of odours. Nothing indeed is more variable than the testimony of each person on this point. Certain odours pleasant to some people offend the nostrils of their fellows.

of pleasantness or unpleasantness, according as the incident excitation has gratified or run counter to their natural sensibility. For this special group of nerves the impression agreeably felt is expressed by the word *perfume*; the impression disagreeably felt by the word *stink*. These are the two extreme terms between which all the shades of their peculiar sensibility are developed. They are incapable of penetrating profoundly into the recesses of our inner sensibility, to excite those grand movements of expansion or depression which are epitomised in the sentiments of joy or sorrow. From this point of view they are very inferior to optic and acoustic impressions, which monopolize the power of exciting the vibrations of the sensitive chords of our human nature. They only excite, then, a limited action of the *sensorium* on their arrival. On the other hand, if their diffusive power does not extend to the emotional sphere, it is reverberated in a very direct manner throughout both the vegetative sphere and that of the natural sensibility of certain points of the *sensorium*, and, when examined from this point of view, olfactory impressions have reflex effects which are quite unexpected.

Thus, we all know that certain odorous substances particularly predispose us to nausea; that certain appetising substances, and the odour of preparations made with vinegar, gum-dragon, etc., act upon the salivary secretion, and, as we say, make our mouths water; that perfumes and certain specific odours have an aphrodisiac action; that with certain impressionable persons the presence of certain odours produces profound disturbances, sometimes even syncope; that finally, in certain persons subject to headaches, it is no longer the *sensorium*

EVOLUTION OF SENSORIAL IMPRESSIONS. 277

as a centre of reception for the moral sensibility that is affected by them, but the sensitive *sensorium*, the brain itself, that is impressed in a painful manner, in certain of its histological elements. Many persons are aware that the odour of certain flowers that make an agreeable impression on their *sensorium* produces a painful after-effect, as though they had to do with a physical ache.

Olfactory excitations are, like their fellows, capable of being stored up in the *sensorium* in the form of persistent reminiscences, and of being associated either with visual impressions or with those sensitive impressions which have been simultaneously imprinted upon us. They are similarly linked with our ideas, and the sentiments that have accompanied their genesis, so that the chance arrival of a perfume in the nostrils, is sufficient to awake a whole series of contemporary memories, and of emotions which arise in consequence, and recall to us the moment and the place in which the perfume was first inhaled.

Olfactory impressions, again, furnish the intellect with precise and specific *data*, which, when preserved in the form of reminiscences and compared together, become materials by means of which we fortify certain judgments.

Thus when associated with their fellows, gustatory excitations, which they perfect and complete in the act of deglutition, they furnish us with precise notions respecting the flavour and sapid qualities of the substances we are eating.

They also warn us, by an act of memory and experience, of the presence of fœtid emanations floating in the air or in the liquids we absorb. They are thus like advanced guards that watch incessantly over the security of the operations of the vegetative life of the human being.

Evolution of Gustatory Impressions.—Collected on the surface of the buccal and lingual mucous membranes, in the terminal expansions of the glosso-pharyngeal and lingual nerves, gustatory impressions are thence probably distributed within a definite region of the optic thalamus; but up to the present time, we are not in a position to demonstrate the precise place of their condensation. From this point they are, like all other impressions, distributed in the cerebral cortex, their area of distribution here also not being yet determined.

1. Intimately connected with their companion olfactory impressions, in their method of impressing the *sensorium*, and being constantly associated with them, they owe to this union a notable portion of their energy, and the various forms in which they reveal themselves in us. Thus it is that the capacity we have for tasting the flavour of certain sapid substances, such as the bouquet of some wines, is only the combined effect of olfactory and gustative impressions, these latter being quite incapable of producing such a result, as we may assure ourselves by stopping our nostrils and allowing our gustatory impressions to act alone. We then perceive how restrained is their field of activity.

They give us the unanalysable and specific notion of sweet, saccharine, salt, acid, acrid, and bitter savours. The diapason of tones that they set vibrating in the *sensorium* is, as we can see, by no means rich in varied shades.

2. *Genesis of the Notion of Good and Evil.*—On the

other hand, they present this very characteristic quality, that the mode in which their extreme notes affect the *sensorium* is so significant and so typical that they constitute for it two quite peculiar and original conditions, which assist us in judging and comparing certain phenomena of the moral order.

Thus, when the natural sensibility of our gustatory nerves has been gratified, when a sapid substance has brought them into a pleasant condition, this peculiar state of satisfaction is transmitted to the *sensorium*, is there propagated, and produces an analogous condition ; and this analogous condition, initiated by the peripheral nerves, becomes a subjective notion, the notion of *goodness*—equivalent to the notion of *beauty* excited in the *sensorium* by the optic nerves when agreeably impressed. We say then that a thing is good when it has fully satisfied our gustatory nerves; so that this peculiar word, primarily applied to the agreeable perception of a sapid substance, is generalized in the *sensorium*, and becomes a moral appreciation which we unconsciously apply to a whole series of acts of the human activity. We declare them *good*, and consider them as *good actions*, merely because they have produced in us, in the emotional regions of our moral sensibility, an impression equivalent to that which a gustatory impression agreeably perceived determines in the *sensorium*.

Inversely, bitter substances, which cause the nerves of taste to shrink, produce in the *sensorium* a disagreeable reverberation, and inevitably become, under the designation of *bad* substances, the expression of a painful impression in opposition to the last, and equivalent to that of pain in the purely sensitive order of phenomena.

This specific notion is thus susceptible of being generalized, of becoming subjective, and of being applied to the appreciation of purely moral actions, which we declare *evil*, tainted with wickedness, because they have, without our knowledge, developed in the *sensorium* a painful impression, equivalent to that produced by a disagreeable gustatory impression.

3. Gustatory impressions, though incapable of causing great shocks in the emotional regions of our personality, are, like their companions, olfactory impressions, capable of radiating into the different regions of the vegetative sphere; they are both of them fundamental excitations of this special division of cerebral life.

Thus it is they which directly regulate the functions of the stomach, and through these the life of the organism. Every one knows what a state of erethism is produced in the gastric mucous membrane by sapid, appetizing substances, and what dulness of appetite is produced by insipid ones; the good appetite produced by the former having a direct influence upon the harmony of the psychic and intellectual activity.

Former gustatory excitations, preserved in the *sensorium* in the form of persistent reminiscences, are on this account easily evoked, and may be compared with recent ones. They are likewise capable of awaking old memories, contemporaneous with the moments in which they have been deposited in the *sensorium*, and of reviving past emotions and the old associations of ideas that have accompanied their genesis. Thus the taste of food, wine, or a liqueur, recalls to us such or such a period of our youth, such or such an episode of our life, such or such an incident in our travels. Thus gustative impressions, like all their fellows, live with the same life that these do, and participate in the same processes of cerebral activity. United to their partners, olfactory impressions, they have a truly specific and penetrating radiation, which extends at once into the domain of intellectual activity and that of purely vegetative life. They thus become the occasion of a series of memories and comparisons, and of the different gastronomic judgments that we form respecting the degree of sapidness of food, the pre-eminence of certain vintages, and the rules respecting alimentary hygiene. They become, when intelligently directed, the occasion of a series of particular satisfactions which are associated with all others, and, as Brillat-Savarin has so well expressed it, outlive all the rest to console us for their loss.

Evolution of Genital Impressions.—Genital excitations, as regards their genesis, their passage through the nervous system, and their diffusion in the *sensorium*, present the most remarkable analogies to gustatory impressions, of which they are to some extent a copy.

Like these, they have no nerves of special sensation; like these they are conducted into the central regions by means of radicle-filaments which are there dispersed according to the special mode of distribution of the posterior roots of general sensibility;* and like these they are distributed to the substance of the central grey matter of the optic thalamus, and then to the plexuses

---

* We know also, that in their centripetal course they are extended, with the conductors which carry them, over the floor of the fourth ventricle, and that lesions of this locality are apt to produce erection, as in those who are hung.
See Luys " Recherches sur le système nerveux," pp. 340, 342.

of the *sensorium*. It is, however, as yet impossible to determine precisely either the special nucleus reserved for them in the optic thalamus, or the territory where their dissemination among the plexuses of the *sensorium* is effected.

Finally, like gustatory impressions, they are intermittent and subordinated to the chance arrival of the causes that determine them ; and, as the last point of analogy, if they are as fugitive they compensate for this by their vividness, their intensity, their suddenness ; by the profound manner in which they affect the *sensorium*, and by the ephemeral character of their manifestations.

Collected principally on the surface of the plexuses of the genital organs which are so rich in erectile papillæ, genital excitations present at the moment of their genesis (in much ampler proportions) that special phase of erethism common to all their fellow excitations, when the sensorial impression radiating from the external world is reverberated in the sensitive plexuses and becomes incarnate in the organism.

For this special order of excitations the primordial period of erethism which incarnates them in the organism, instead of being a local and instantaneous phenomenon like those of general sensibility, or vision, for instance, is divided into successive moments. It is effected by means of special erectile apparatuses, which develop, and complete it, and insensibly lead up to a condition of supreme exaltation. Once the external impression is incarnated in the sensitive plexuses, once the notion of physical pleasure is developed with all its consequences, it ceases to be itself, through dynamic

EVOLUTION OF SENSORIAL IMPRESSIONS. 283

exhaustion, sheer fatigue of the nerves, as we have seen the retina when fatigued becomes insensible to the contemplation of certain luminous rays.

The process of physical pleasure undergoes, then, a series of phases through which it only gradually arrives at its complete expansion.

It begins locally, in the peripheral plexuses, with a period of extreme erethism, from the intimate connection of the sexes, through the mysterious conjunctions of the apparatuses of organic life; it is at the same time enriched by the action and sympathetic participation of all the diffuse sensibilities of the organism which are thrown into agitation, those of the tactile surfaces, the hands, the lips, which all combine to enhance its primitive energy; it advances towards the central regions, as a true synthesis of all the impressionable elements of our nature in vibration, is propagated through the whole length of the spinal axis by means of the conducting fibres which convey it, and, after passing through its final stages in the intermediate grey regions of the optic thalamus, it is dispersed in the different zones of the *sensorium*, carrying with it the shock of joy and satisfaction which intrinsically characterizes it.

Like all the other sensorial impressions, the excitations of physical pleasure affect both the sphere of psychical activity and that of intellectual activity proper.

1. The excitations of physical pleasure, which, as regards the living being, represent the fundamental elements of the prime function which has for its end the reproduction of the species, arrive in the *sensorium*

accompanied by an enormous contingent of sensations emanating from different regions simultaneously in a condition of erethism. They essentially carry with them impressions of joy and happiness, and produce like conditions in the elements of the *sensorium;* becoming during the period of puberty a dominant note which vibrates above all the rest, which gives its tone to all our actions, all our discourses; and which, when it happens to be set vibrating with special intensity, extinguishes all the rest by its intensity and splendour.

Psychic, ideal love, and physical love are, then, the ultimate links of one and the same chain of which the elements are uninterruptedly connected. It is a regular physiological process, which has its roots in the intimate connection of the sexes, and its expansion in the most elevated regions of psycho-intellectual activity. In evolving itself throughout the organism, it thus involves the incidental calling into play of all the apparatuses of the essential life of the living creature, and their harmonious co-operation.

It has, then, its *raison d'être* in a purely physical pleasurable excitation, which presides over its origin and marks its first stage. It is a fleeting and transient desire, which is born, passes, and fades away as soon as the physical demands for pleasure which gave it birth are appeased; but as the same physical needs arise again, through the necessary laws of the movement of life in living beings, the same voluptuous desires simultaneously arising in the *sensorium*, it follows that the reiteration of the same physical satisfactions finally leaves upon the *sensorium* itself a persistent and continuous impression, vibrating like an echo of the past, and thus maintaining a durable and uninterrupted sentiment. Thus it is that love, a sentiment transient and ephemeral as the pleasure which gave it birth, fixes itself permanently, and lives with a life of its own. The reiteration of the satisfaction of physical pleasure, obtained from the same sources as formerly, and new desires resuscitate and reinvigorate it, and become the elements of its continuity and its persistence.

Conjugal love, thus made an abiding sentiment in the *sensorium*, becomes in its turn the physiological pivot around which a new generation of consecutive sentiments gravitates.

Thus, by the natural fact of the evolution of the living organism, physical love, which was at first all concentrated upon a single head, upon the being which gave it birth—its end being the propagation of the species—when once this end is attained is insensibly extended to the offspring, which is the flesh of the flesh of this being, and the veritable proliferation of her substance. The sentiments of the family which are then developed, lead man's emotional nature into the inevitable cycle of the affection of parents for their children, that inevitable cycle in which we have been preceded by all the generations of our ancestors, and in which all the representatives of the human race are destined perpetually to move.

Here the process of physical love finds its last stage, dying out of itself after it has accomplished its work, by developing in the living creature, during the period of his maturity, all the energies of his organization, animating his heart with the most intense emotions, inspiring the liveliest sallies of his intellect and imagination, and contributing necessarily to the perpetuation of his race and the preservation of his species.

2. In the domain of intellectual activity proper, the excitations of physical pleasure have an action as powerful as that they exercise in the sphere of purely psychical phenomena.

In proportion as the human being who has passed through the transitory phases of puberty accomplishes his physiological evolution, new ideas arise, unappeased desires are awakened; he feels himself incomplete in his solitude, and comprehends that another being is designed to fill the void of his sentiments and desires.

Henceforth, urged on by his latent desires, he employs all the resources of his intelligence to seek out his future companion and to prepare for her the necessary material provision. He thinks of his social establishment; he struggles ardently in the battle of life: the woman, and union with her in marriage, are the secret motives of his actions. It is the hope of attaining this end that sustains his strength and maintains his courage; and, later on, when he has attained this end, he still struggles (and puts forth all his intellectual activity in the struggle) to save his offspring from the troubles of the road along which they must follow him. He thinks of the future, and prepares the inheritance he will leave behind. He thus harmonizes all the intellectual activities, all the social forces he can command, with the different phases of the physiological process which is being inevitably worked out in him; and under the most diverse forms, in the most dissimilar circumstances, he always obeys the same necessary laws of evolution that press upon him and metamorphose him insensibly, from the moment in which he becomes a candidate for marriage to that in which, after having been a husband and father, he becomes a grandfather, and sees in the second generation that springs up around him the secondary ramifications of the branches of which he is the parent stem. So that, whatever be the position of a man (I mean of a complete and regularly constituted man), on whatever rung of the social ladder we may imagine him placed, we are always sure to find at the bottom of his actions, open or secret, as the first cause of their motives, the craving for physical pleasure, and as a consequence, psychic pleasure, with all the sentiments to which it gives birth. It is this which, always present, always active, becomes in every act of his life the natural stimulus of the briskness of his mind, the resources of his imagination, and the vigour with which he enters upon the struggle for existence. It tinges his whole personality, animates him incessantly, and produces such concordant action of all his powers that we may say, without fear of mistake, that the measure of his physical is also that of his moral virility.

3. Genital excitations play such an important part in the sum total of the operations of psycho-intellectual life, that when they are arrested in their development, in consequence of certain operations that nip them in the bud in the regions where they have their point of origin, a very remarkable effect is produced upon the intellect and character.

Every one knows how mild and easy castrated animals are to manage, and how this fits them for the rule of man, through the modification of their natural impetuosity. In man, the same practice produces similar effects. According to Godard,* castration performed on the adult singularly weakens the moral energy, as the following fact, reported by d'Escayrac, de Lauture, proves. "I have seen," he says, "six slaves belonging to the kachef of Abouharas, in Kordofan, who, in consequence of a conspiracy against the life of their master, were emasculated by him. All were adults at the time of this mutilation, and none of them died. Their characters changed completely, and the submission they now show differs remarkably from the spirit of rebellion that animated them previously."

Godard afterwards adds† that, according to Dionis, castrated persons are unsociable, liars, and rascals, and that they never seem to practise any human virtue; and that, according to Benoît Mojou, eunuchs are the vilest class of the human race, cowards and rascals because they are weak, envious and spiteful because they are unhappy.

Finally, he has noticed that even where no mutilation has been practised, individuals with congenital absence of the two testicles are effeminate, unenergetic, timid; they blush easily, everything frightens them, and it is difficult even to examine them without a great deal of trouble.

* Godard, "Recherches tératologiques sur l'apoareil séminal de l'homme," p. 68. · Paris, 1860.
† Loco citato, p. 73.

CHAPTER III.

THE JUDGMENT.

JUDGMENT is the principal operation of cerebral activity, by means of which the human personality, in presence of an excitation from the external world, either physical or moral, expresses its condition.

. Among the diverse operations of the brain in action, that of judging is a regular physiological process, which is developed according to fixed laws and inevitable organic conditions, and which, like the different phenomena of muscular activity (the progression of the human body in space, for instance), expresses life in exercise and the nervous power in a dynamic state.

The action of judging, so far as it is a physiological process accomplished by means of the cerebral activities in movement, is decomposable into three phases, which are as follows:—

1. A phase of incidence, during which the external excitation impresses the *sensorium* and rouses the conscious personality to action.

2. An intermediate phase during which the personality, seized upon and impressed, develops its latent capacities, and reacts in a specific manner.

3. A final phase of reflexion, during which the process, continuing its progress through the cerebral tissue,

is projected outwards in phonetic or written co-ordinated manifestations. The impressed human personality, in fact, expresses itself, exhales itself in its entirety, in either articulate or written language.

1. It is always a recent or former sensorial impression that naturally excites an operation of the judgment and determines its action. The *sensorium* is impressed, the human personality takes part in the phenomenon; it is strongly affected, and reacts immediately. This work of absorption of the sensorial excitation and of *conscious* reaction, on the part of the personality, implies then a series of connected operations which follow and complete one another, like the different phases of a simple somatic process. It even requires a certain appreciable time, to be effected in the cerebral tissue, and, according to the nature of the individual, will act with greater or less facility, and perfect itself with exercise, as Donders has demonstrated.*

* Donders, by means of very ingenious registering instruments, has succeeded in introducing a precise notation in studying the evolution of certain phenomena of the cerebral activity. The method consists in making an impression upon a person and noting the precise instant at which he responds to it. The person who makes the experiment must, as soon as the impression is felt, press with his finger a spring which sets a revolving cylinder in motion. The number of revolutions indicates the time that has elapsed, that is to say the time necessary to permit the complete process of the judgment, the impregnation of the *sensorium* and its expressed reaction, to manifest themselves externally. The precise duration of voluntary transmission is known, since it is always pretty much the same, and thus we arrive at the knowledge that a luminous sensation is more quickly perceived than an acoustic or a tactic. In this case it is a simple thought that is transmitted.

Donders again applied himself to ascertain by the same process the time necessary to solve a dilemma. A person is in darkness, a green or red light is flashed upon him, and he is to make a certain signal with the right or left hand according to the colour exhibited. The sum of these operations is more complex and requires much more time; but, as the elements of the previous experiment are here again found, we have only to deduct the time necessary for this,

It is in this first phase of the operation that the whole secret of its final rectitude resides; for *to see well* and *to judge well* are synonymous, and to acquire the power of pronouncing with certainty, respecting such or such a circumstance, we cannot surround ourselves with too many precautions.

Nothing, in fact, is more difficult than to have a clear and precise appreciation of real things. The minute care taken by physicists and chemists, and the infinite precautions with which they surround themselves, in order to appreciate simple physical phenomena, show us how frequent are the causes of error, and how liable to deception is all observation; since we so often find two observers, in the presence of the same physical and palpable phenomenon, each describing it in his own fashion, and each giving a very different report respecting it.

A fortiori we can understand that when we have to do with the interpretation of complex things, to form judgments respecting history, contemporaneous or past; respecting the facts of our current life, in which all human passions are openly or secretly at work; respecting political matters; the ascertainment of the real facts may become very difficult, the very notion of truth obscure. We see how those judgments, which we succeed in formulating, always fail at some point or another, from the intervention, more or less eager, of our own personality.

---

to ascertain the time required by the brain to discern whether the light was green or red, and which hand was to be used. Donders, "Archives néerlandaises," 1867, vol. ii. Instrument for measuring the time necessary for psychical acts.

2. The second phase of the process is no less delicate than the first; for here the human personality, on the advent of more or less clearly distinguishable stimuli from the external world, comes into play with all its sensibilities awake, reacting, like a trustworthy reagent, when the excitable regions of its inmost core have been more or less affected.

It is the human personality that *feels*, that is *moved*, that speaks in our judgments, and that reacts in an appropriate manner, according as it is restless, impressionable, indifferent, or atonic; reflecting externally in words or deeds, the infinite varieties of feeling that lie maturing in its recesses. Like a true *leading-note*, it vibrates every instant in every act of our lives, and gives our judgments an original character according to the key in which it is pitched, a something racy of the soil, which (when once our *amour propre* comes into play, and our own personality is concerned) always expresses the different phases through which our *sensorium* passes when in a state of agitation.

Hence, the difficulty of forming impartial judgments in questions of the moral kind, the judges being biassed; hence, that series of minute precautions taken by legislators at every step, to eliminate interested persons from juries, and to form these of independent individuals free from all prejudices. Hence, that practical observation, verified by every day's experience, that young and ardent natures in whom the effervescence of the *sensorium* is still unabated, are apt to judge of men and things with all the rapidity and prejudice of their characters; and that the judgment is exercised in a more enlightened manner when maturity has arrived, and the wear and tear of life have exhausted the first ardours of the natural sensibility. Cold contemplation of the real facts is more easily attained, and permits the human personality to expand in a calmer and more reflecting manner.

It is therefore in this intermediate phase of its evolution, when it enters into contact with the human personality, that the process which is destined to be converted into judgment comes to its crisis, according to the variable emotivity of the substratum that receives it.

When the phenomenon is produced, two circumstances may occur: either the process may achieve its evolution, and appear externally in a verbal or manuscript formula which epitomizes it; or it may die out on the spot, remain silent, and, like a living force which undergoes transformation, may proceed to excite secondary impressions throughout the cerebral regions it traverses. New territories of affected cells will then come into play, and according to their automatic activity will associate themselves with the excitations and ideas in question. Thus it is that a process of judgment, suspended in its course, becomes the local origin of a vibratory movement which radiates to a distance and produces secondary impressions. It is because of this physiological radiation that related ideas are automatically excited; that new views arise, manners of looking at the matter not at first dreamed of; and that, from this work of internal digestion of the process in evolution, a whole series of new considerations springs up and gives to the first judgment a weight it had not before, and the natural complements of its real value.

The process of judgment has then for its special

characteristic, according as it advances, the privilege of extending itself; of determining the reaction of the surrounding cerebral elements; of searching, to some extent, into the archives of the past; of associating former notions with those of the present; of creating partial local judgments, established *à priori* as results of the inner experience of the individual; and of permitting us, at a given moment, to juxtapose and agglomerate partial judgments—to *agglutinate* them, in the form of arguments, into a complete judgment, which resumes them all in a true synthesis.

Thus, for instance, when I auscultate the chest of a patient, and perceiving the existence of tubular respiration, declare that the patient is in the second stage of pneumonia, I give utterance to a judgment that has many ramifications in my mind, and is made up of a great number of different materials. Starting from this blowing noise that has struck my ear, I represent to myself what, under similar circumstances, I have perceived on previous occasions. I have observed, for instance, that this blowing noise corresponds to a hyperæmia of the pulmonary tissue, with concomitant induration, that it depends upon an induration of tissue, not upon the presence of effused fluid. At the same time I perceive with my eyes the general condition of the patient, I note his countenance, his external habit, the state of his tongue, etc., and a new series of notions acquired by the exercise of optic impressions is awakened in my mind and becomes associated with the process already begun by the auditory impressions. I percuss, moreover; I feel the pulse; I palpate; and once more, starting from a new series of sensorial impres-

THE JUDGMENT. 295

sions that come into play, new regions of the *sensorium* are associated, set in vibration, and take their part in the complex operation that is taking place. The different regions of my brain are successively affected. Notions formerly acquired are laid under contribution; they come forward of their own accord on the occurrence of the excitation with which they are methodically connected as contemporary memories; and thus the personality, reminded of the primordial impression, and enlightened by the total product of the related notions that spring up automatically, pronounces its judgment with a sufficient number of materials, and expresses the manner in which it is effected in a verbal form which is the index of its present condition. Thus it is that in pronouncing the words "pneumonia—second stage," I epitomize a whole series of former notions, methodically grouped, which have made their appearance in my mind *motu proprio*.

Natural Predispositions.—In this second phase of the cerebral process, which is being accomplished, the human personality is seized on, as we have said, and inevitably associated in its evolution. Here a new peculiarity, which occupies an important place in the phenomena of cerebral life, comes in; viz., the manner in which that personality is brought into play and the particular mode in which the sensorial excitation has affected it.

We have already insisted (p. 43) upon the curious relations that exist between the different provinces of the cortical substance and certain centres of the optic thalami with which they are more particularly connected. We have thus shown that such or such a group of sensorial impressions was more especially distributed to such or such a region of the cerebral cortex; and we have at the same time made it clear to what an extent the greater or less richness in cells of such or such a cerebral region, and the briskness and impressionability of these cells themselves, may induce certain functional predominances, and become the natural cause of certain dispositions and special aptitudes of the mind.

In applying these data to the evolution of the process of the judgment, we recognize the fact that if the human personality, at the moment it begins to take part in this, finds in one of the regions of excitation a greater number of nervous elements than in such or such another; if the elements are more impressionable, more vivacious, better co-ordinated in their internal mechanism, it will be on this account more strongly impressed, and provided with means of expression more rich and more abundant.

Thus, served by the best instruments, it will react in a more complete manner; will do what others, less richly endowed, could not do; will see better, hear better, taste better, smell better, etc. It is by means of these natural conditions of organization that certain individuals show themselves superior to others as regards the operations of the judgment, in the direct ratio of the superiority of their cerebral constitution.

On the other hand it is notorious that, just as all the sensorial organs are not gifted with the same energies in all individuals, and that one is marvellously gifted for music, another for drawing, another for painting, etc., so by reason of that pre-eminence of certain impressions in the *sensorium*, which constitutes in a manner the cerebral temperament of the individual, it results that in the total of mental faculties whatever cerebral region is best furnished, will be the privileged region in whatever operations of the judgment are the best and most rapidly accomplished. Hence will also arise partially competent judgments, the individual being better fitted to judge pertinently respecting some one particular subject. Hence, according to our individualities, those striking contrasts of which we daily see so many examples, where we meet with persons who judge soundly respecting some subject they have thoroughly studied, or which is their "hobby," who are yet completely incapable of forming an ordinary judgment respecting a simple question of everyday life. The human mind, limited in its resources, and the tributary of the nervous elements through whose instrumentality it manifests itself, is only capable of isolated and restrained efforts; and thus it is that in the infinite variety of its manifestations we see what a division of labour man must adopt to concentrate his energies upon a point so as to bring them to bear with regularity, and, in a word, how truly judgment is said to be a most difficult operation —*judicium difficile.**

3. The process of judgment, when once it has called

* It is strange to observe how often, within pathological limits, we meet with individuals who partially preserve their capacity for judging of certain things. We see in fact lunatics who can sustain a connected conversation, provided that those points which bring their personality into play be avoided. If we accidentally touch the sensitive chord the dissonance suddenly bursts out and the delirious conception becomes clear. There are others who are completely incapable of judging of things around them, of acting with discernment where their own interests are concerned, and who notwithstanding preserve an aptitude for certain games, which require that the attention shall be sustained over a limited field, the game of draughts, for instance, which demands the contemplation of the draughtboard, without necessitating efforts of memory, as games of cards do.

forth, as it passes, the participation of the different regions of the cerebral cortex, and has associated itself with the human personality, tends more and more to effect its extrinsic manifestation, and to express itself outwardly either in suitable articulate sounds, by which custom has taught us to express the different shades of our sensibility, or in the form of graphic characters which similarly signify our ideas and inmost thoughts.

Henceforth it assumes in the *sensorium* the form of a *conscious resolution*, and, from this moment, the spontaneous voluntary act is similarly completed in its essential elements; since the cerebral operation in which it is essentially embodied, the awakening of the human personality, *conscious* of what is taking place, has occurred, and is about to reveal itself externally under the most diverse forms. From this moment the process of judgment, in its third phase, belongs to the series of the phenomena of voluntary activity, of which it marks the first stage. It then embodies itself in the somatic translation of a voluntary excitation radiating from the psycho-intellectual regions. We shall now follow it in this last phase, by explaining the action of voluntary motor-power.

Community and Points of Contact of Human Judgments.—Common Sense.—Once, now, the process of the judgment has been externally manifested, and by this has become capable of implanting itself in the brain of another and determining in him similar reactions,— once, I say, this operation has been accomplished, how is it possible to appreciate exactly the value of the physiological act that has been effected? How can we discern the justice of the opinions arrived at, and know whether the judgment formulated be true or false, and, as we say, reasonable or unreasonable?

When dealing with the discernment of things which fall immediately within the domain of intellectual activity, it is comparatively easy for each of us to know that a judgment pronounced is conformable with truth and reason.

Every one knows that in the domain of science, all the fundamental truths which are the common patrimony of the human mind, in evolution from century to century, be they mathematical, chemical, physical or biological, are universally accepted; that what is true in Paris in astronomy is similarly true in Pekin or New York; and that in all places in the world, wherever they meet with a sensible and well-informed man, they are well-received and comprehended.

Now this universal concord, this acquiescence of all in their acceptance as legitimate and truthful judgments, exists because they only express evident and precise ideas, verifiable by experience; because every one can directly or indirectly put them to the test; and because the human personality that observed and expressed them for the first time had nothing to do with their genesis, except the expressing of them in correct and appropriate terms, the emotional regions of the sensibility not having been laid under contribution in the smallest degree.

The real only, and nothing but the real, is revealed in the exposition of each of them; and the individual who has expressed them, having perceived the external world in an incident form, has but reflected them externally without adding anything of his own.

Thus, when Copernicus or Kepler formulated his laws of the system of the world and the movements of the planets; when Newton made evident the decomposition of light into its elementary rays; when Lavoisier demonstrated the part played by oxygen in the phenomena of combustion and respiration; when Laennec furnished his contemporaries with a new means of penetrating with the ear the machinery of the living human frame, and following step by step the respiratory movements and those of the heart,—these were new truths, unexpected judgments that were thrown into the intellectual domain, and which, as a correct expression of reality, and certified as conformable to this by every one interested, were addressed to but one region of the living organism, the intellectual, without being addressed to the emotional regions, and without exciting the slightest passion. These are palpable, tangible, verifiable judgments, which, being addressed to all, true for the future as for the present, present those general characters proper to grand truths, permanence and universality.

If it be generally possible to appreciate the regularity of a process of the judgment in the sphere of purely intellectual phenomena, by mediate or immediate verification, it, on the contrary, becomes very difficult when we have to judge of a question which belongs to the class of moral phenomena.

Here all becomes complicated and obscure; for the criterion of verification, experience, which we had before, is here wanting. There is no standard by which to measure the things of the moral order; this incident, fact, or particular document which has to be judged of, from the mere fact of being a direct emanation from some one else's personality, his private opinion which is externally revealed, borrows from the emotional regions whence it proceeds a specific colouring; his private personality is more or less at work, with its emotions and passions.

On the other hand, we ourselves, who have to judge of this incident, this document, these words, are similarly unconsciously affected by latent sympathies or antipathies, which make us see and judge of the thing under colours which are not always those of reality.

We see, then, of what multiple elements the action of judging of a phenomenon of the moral class is composed, and how many unforeseen factors, variable at every instant according to the state of our natural sensibility, come in at cross purposes to drive us away from the desired goal.

Thus, in the special domain in which moral sensibility reigns alone, we may say that the experimental methods of valuation are entirely at fault. We must, therefore, have recourse to entirely new methods, considerations of a moral kind which shall serve as a common measure, and which, when applied to the valuation of phenomena of the same nature, may be capable of leading us to a solution of the problem, and the formation of a judgment respecting its nature.

If it be true, indeed, that in human practice and the ordinary affairs of everyday life, there is nothing that differs so much from one man as another man (since we each carry in us the weight of hereditary influences, influences of race and education accumulated through long periods of time, and the shades of sensibility of each of us are as different as the details of our persons), there is, nevertheless, in that sum total of data which constitute the elements of the moral life of man, a common stock of *fundamental truths* which form, as it were, a series of moral axioms and a veritable patrimony, proper to all sentient humanity. In all times, and everywhere, indeed, it has always been a fine thing for a man to serve his country, to sacrifice himself for his kind, to honour his parents, to bring up his family well, or, to make use of a formula which contains an epitome of universal morality, to do or not to do to others as we should like others to do or not to do to us, etc. Within a more restrained circle of ideas, we know that in unions of men agglomerated into isolated societies, though they be independent or even enemies, there is a common fund of ideas and sentiments. Among soldiers, under whatever flag they serve, the sentiment of military honour is always the same. The *esprit de corps*, which we see developed in certain associations, is nothing but the resultant of a community of ideas and sentiments among all the individuals living in society, and united by the bonds of a vast confraternity.

In all times and places, then, this collection of common ideas and sentiments which serves as a basis for phenomena of the moral order, has been, as it were, a sort of directing clue for humanity, a *magnetic meridian* of common sympathy, by which men have unconsciously regulated their conduct ; and this is so true, this common fund of moral sensibility is so inherent in our natural sensibility, in our very personality ; it is so vivid in us, and so organically constituted, that wherever we find one of our fellow-creatures we judge, *à priori*, that he must vibrate in the same keys, and thrill to the same impressions. In a word, we believe in the existence of this moral sensibility in others, with the same certainty that we feel regarding the existence of his heart that beats, his lungs that breathe, and his limbs that move according to flexions and extensions previously determined.

This common basis of moral sensibility which lives within us and extends to all our fellow-creatures, forming a bond of universal sympathy between all members of the human family, thus becomes the veritable *criterion* and touchstone that serves us to appreciate and judge of the value of a phenomenon of the moral kind. To a particular phenomenon we logically apply a particular method of diagnosis. It is by taking ourselves as a term of comparison, by bringing our conscious personality into the presence of the actions of another, by placing ourselves in imagination in his place, that we arrive at a notion of their scope, and a judgment as to whether they are conformable to the common average-line of human sentiments and universal sensibility.

We thus arrive at the conclusion that there are among mankind fundamental truths of the moral kind, common modes of feeling, which we all unconsciously obey, and which constitute the *common line of average*, the *common sense*, according to which the great human family advances along the path of life. Each of us takes the bearing of his acts more or less from this, and, if these deviate from it, this deviation is then felt by those who are following it, and they accordingly judge of it and condemn it, as a deviation from the common law, and as the patent expression of a perturbation which has occurred in the faculties of him who has thus got out of the common rut.

We accordingly consider every word, and every piece of writing that is understood and accepted by all, *reasonable*, according to *common sense*; while on the other hand, we characterize as *unreasonable* every action that shocks the notion of right sense and rectitude of judgment, as they exist in others.

Thus, that conception of things in their totality, which we designate under the term *reason*, is generally, from a physiological point of view, nothing but an abstract synthetic expression which serves to express that unconscious tendency we have to follow, in our lives, our ideas, and our actions, the common course followed by our kind, and not to deviate from the meridian line followed by the majority.

Functional Perturbations of Operations of the Judgment.—A study of the morbid forms of the operations of the judgment, shows us how closely united one with another are the different phenomena of which it is constituted, and to what an extent the whole becomes perturbed and disordered, when one of these comes to be disturbed in its mode of action, (especially the first, which is the most important, and the point of departure of the operation which takes place); and how far the external expression which results, is in more or less complete discord with the reality of things.

The first phase corresponds, as we have said, to the moment in which the external impression penetrates the *sensorium*, and seizes upon the personality, which immediately participates in the communicated impression. This is the delicate moment of the process, when the terms of the problem are stated. Now, what happens when this primordial sensation which should arrive at the *sensorium* with the maximum of precision, and reflect, in as exact a manner as possible, the surrounding phenomena, is incompletely transmitted and falsified; when, from some accidental disturbance in the different centripetal apparatuses charged with its collection and transmission to the *sensorium*, it arrives there deprived of its essential character and incompletely expressed (sensorial illusions)? What happens when, on the other hand, the intermediate regions, whose mission it is to transmit to the *sensorium* peripheral excitations (centres of the optic thalamus), assume a condition of automatic erethism, and proceed, *motu proprio*, to launch towards the *sensorium* subjective excitations engendered on the spot (hallucinations)?

The human personality, then without any means of direct control, seized upon by fictitious autogenous excitations, according to natural processes, accepts the change; receives them, absorbs them, works them up, submits them to the same subtle operations as though they were the regular and legitimate aliments of its activity; and henceforward the abnormal process, by means of the working of the energies proper to the cerebral elements, and by virtue of habits formerly acquired, goes on of itself, as logically and inevitably as though it were a pure emanation from the real world; of course, to the great stupefaction of persons who are not initiated into the knowledge of mental diseases, and cannot bring themselves to admit that a false conclusion may be deduced with perfect logic, and that logic does not imply either the justice or the precision of any judgment whatsoever.

Whether the protopathic excitation, then, be regularly or irregularly engendered, all goes on in the brain automatically and, to some extent, unconsciously, by the individual force of the organs traversed by the process in evolution; as though we had to do with a simple reflex operation in process of development in the grey tissue of the medulla; as though we had to do with a foreign body, or a poisonous substance accidentally introduced into the stomach, and inevitably passing on its way through the successive regions of the intestinal canal.

We can thus comprehend how the third phase of the process (which is but the ultimate expression of the period of apparition and exteriorization of the human personality, which manifests its peculiar emotivity) expresses, in a corresponding manner, the different vices of organization that have accompanied the first moments of its genesis.

In fact, if we study the concatenation of ideas and arguments in the case of the most rational lunatics, in those who, with persuasive logic, express, in correct terms, and often in a winning and convincing manner, all their emotions and all their extravagant conceptions; if we follow out with care the natural sequence of their wanderings, we shall always find that the first origin of their arguments and recriminations, their ideas of those persecutions of which they accuse those who surround them, their family, society in general, or persons undefined, have for their primary point of departure, an initial disturbance occurring in the method of sensorial perception, and in the initiatory phase of a process of judgment.

It is always a sensorial illusion, an hallucination, that is at the bottom of the morbid act, and directs its inevitable course.

Thus, we sometimes find an energetic and intelligent patient affected with reasoning mania, who bitterly complains of the soiled linen that is given him. He violently attacks those in his service, and complains of the tricks of which he is the victim; then shows the linen objected to, and, lo! it is perfectly clean. We have caught the sensorial illusion causing the extravagant judgment in the moment of its genesis. The patient imagined that he saw a spot of dirt, where there was none; his senses used him badly; and hence a series of extravagances constantly renewed in the same mind abused by its senses, and recurring by means of the same mechanism.

Or again, we may find another who, also suffering from peripheral disturbances in his nervous system—a special condition of his gustatory sensibility—concludes that the food he is given is bad, that powders are put into it, that they wish to poison him, and that such and such a person is guilty. Another has scarcely risen from table, when he makes a great outcry, because, as he complains, they have given him no dinner. He is examined, and it is found that he suffers from a temporary anæsthesia of the pharyngeal mucous membrane.

A laundress, whose case is reported by Charbeyron, and who had given up her business and become a sempstress on account of rheumatic pains, used to work late at night, and got ophthalmia. She, however, continued to work, and saw at the same time four hands, four needles, four seams. She had, in fact, double diplopia. She at first treated this as an hallucination; but, at the end of some days, in consequence of weakness and prolonged mental anxiety, she imagined that she was really sewing four seams at once, and that God, touched by her misfortunes, had worked a miracle in her favour.* As we see here, also, there was a primary disturbance occurring in the first stage of the process (a sensorial illusion, diplopia) determining as its consequence the extravagance and error of judgment.

In other circumstances, there are true hallucinations, phenomena engendered on the spot by a species of erethism of the sensorial channels, which interpose and produce changes in the conscious personality. There are, in fact, almost always hallucinations of hearing, sight, and smell, which, either isolatedly or simultaneously, impinge upon the *sensorium*, and which are almost always found at the bottom of all forms of delirium. Sometimes there are voices heard subjectively, which incite the person under hallucination to avoid such or such a person, or to commit such or such an action; that speak to him in a tone of menace and trouble him in his nightly rest. Sometimes there are various visions which keep him awake, painful perceptions, either of taste or smell, which cause him to refuse food, etc.

Hence an indefinite series of consecutive judgments

* See analogous cases cited by Parchappe, "Annales Médico-psychol.," 1861, p. 271.

THE JUDGMENT.

and reflections, varying infinitely according to the nature of the soil in which they are evolved; hence all those forms of delirium by which the emotions of the personality reveal themselves, and which all have this common basis which unites them one with another, that the morbid conception implanted in the mind as a homogeneous element, and to some extent as a conception contrary to nature, only reveals itself outwardly in a vague and cloudy manner, yet logically, notwithstanding. The person under hallucination, who has vaguely conceived a suspicion in consequence of a low auditory impression which has affected his *sensorium*, outwardly expresses this state of indecision and vague information in the same vague manner; and in this we still find the ordinary methods according to which the processes of the judgment manifest themselves in us. The person under hallucination is vague in his expressions, because the impression which excites his personality is similarly vague and confused. He does not clearly express what he does not clearly understand. He uses only indistinct formulæ to express the conceptions that pass through his mind, always impersonal phrases;—*some one* has told him so and so; *some one* has warned him of so and so; his expressions never being descriptive nor vivid, nor possessed of those distinct outlines that characterize impressions really seen and really heard.

Thus, in fine, we see to what an extent the morbid processes of *divagation*, however wide apart be the different forms they assume, obey the same general laws as the regular processes of the judgment. They pass through the same phases in their operation, by means of the same automatic machinery; they follow logically the same routes; and when they are at discord with reality, when, in a word, their operation has failed, it is because it was badly prepared as regards the arrival of the sensorial impression, and because the phenomena of perception have been disturbed in their essential connections. The human personality, carried away into this fatal cycle, obeys automatically, and inevitably becomes involved in the pathological disorders that occur in the *sensorium*. It is incapable of resisting the strain; and when it comes to its senses, and the disease is cured, it is rather owing to a calming down of the regions primarily affected, than to any action of the conscious volition. The mental condition improves with the physical, and if the divagation disappears, and the individual ceases to be delirious, it is less by a spontaneous effort of his will, by virtue of which he abjures his false convictions and yields to the judgment of others, than because his brain becomes permeable by the surrounding reality, and because he absorbs sensorial impressions, and elaborates them as the generality of mankind do.

We know, indeed, how refractory to all sane reason are men with false ideas, and what a waste of labour it is to endeavour to treat a partially delirious individual by means of logical reasoning.

BOOK III.

PHASE OF REFLECTION OR EMISSION OF THE PROCESSES OF CEREBRAL ACTIVITY.

Preparatory Period. Motor Processes.—In the statement we have just made, we have seen that the processes of cerebral activity, which consist first of all in an impression upon the *sensorium* of external origin, resolve themselves into various reactions on the part of the cerebral apparatuses which are roused into activity, and into a sort of intra-cerebral radiation of the exciting movement.

Now this impression, which has arrived in the form of an incident excitation, is a living force in act of transformation; this force is implanted in the *sensorium*; it becomes reinforced and concentrated according as it is evolved; it is necessary that it shall still continue in motion, and that, under one form or another, it shall pass out of the organism, by discharging itself upon other organs designed to serve it as gates of exit.

From this new stand-point we shall henceforward consider the phenomena of cerebral activity, at the moment in which, in their third phase of evolution, they finish their last stage and reveal themselves in various reactions. These, however varied their appearances, nevertheless represent in the external world the reverberation of a former sensorial impression emanating from this external world.

Once upon their outward course, the processes of cerebral activity take two different routes, according to the variable conditions of receptivity of the cerebral medium in which they are developed, the nature of the individual, and his manner of feeling.

Thus they are sometimes reflected towards the different departments of vegetative life. They do not make their exit from the organism, and in that special sphere they produce secondary commotions of a more or less apparent kind; their reflection takes place in an entirely automatic manner, and in spite of voluntary action (return shock of mental emotions upon the physical constitution).

Sometimes, on the contrary, they appear externally, and reveal themselves by the help of various means of expression—phonetic sounds, graphic signs, appropriate gestures. The external sensorial excitation, radiating from the external world that gave it birth, is in this case directly returned to this external world.

CHAPTER I.

REFLEXION OF MOTOR PROCESSES UPON THE PHENOMENA OF VEGETATIVE LIFE.

IN the first series of facts, when the excitations derived from the external world are not directly reflected outwards—when, under the influence of one cause or another, the primary impression remains confined within our own organism, it dies away there, and the reverberation which results extends to a greater or lesser distance. The nervous discharge of the process, arrested in its course, reacts upon one region or another of vegetative life, and this depends upon the closeness of the sympathetic links uniting each of these with the *sensorium*.

We have shown, on the other hand, that by reason of these connections, there exist, as it were, incessantly permeable natural channels, by which the impressions of the *sensorium* may at any moment become associated with the phenomena of vegetative life, and reverberate throughout the whole extent of the life of the viscera.

The result of this arrangement is that every external excitation arriving in the *sensorium* is sympathetically felt in the different centres of visceral life, and that the slightest excitations that wrinkle the surface of its plexuses, as well as the shocks that overwhelm it, are sympathetically propagated into such or such a department of organic life; now here and now there, centrifugal currents arise instantaneously, and carry to a distance without our knowledge or voluntary participation, prolonged reverberations of the oscillations of the psycho-intellectual sphere.

We all know what an effect painful emotions have upon the phenomena of the circulation; how the heart palpitates without our knowledge when our emotions are at work; how apt this latent over-excitement is to fatigue the vital energy, and what a serious, and long ago recognized influence mental causes have as regards the genesis of its organic lesions; how susceptible the vasomotor innervation is of becoming associated with our emotions in a similar manner; since instantaneous paralysis of the capillaries, on the one hand, is apt to determine those sudden blushes which by showing themselves upon our faces reveal so well, in spite of us, the secrets of our agitated sensibility; while, on the other hand, their spasmodic contraction excites those instantaneous pallors which as directly reflect the perturbations that traverse our *sensorium.*

We all know, moreover, how directly the digestive organs are associated with the impressions of this same *sensorium.* The stomach in particular is intimately connected with the phenomena of cerebral activity. Like the heart, it every instant experiences the return shock of our emotions, and like it, becomes the bearer of the sins of our general sensibility. Every one knows that digestion is disturbed by mental emotions; that vomiting frequently accompanies cerebral disease; and that in certain localized pains of the *sensorium* (hemicrania), when too strong an external excitation evokes its sensibility, the discharge of the *sensorium* in erethism takes effect upon the stomach, which to some extent serves as a gate of exit for the nervous over-excitement reflected towards the organs of vegetative life.

We all know, further, how intimate is the association between the respiratory organs and our natural emotions. Sighs, spasms, anxieties, the involuntary laugh which sometimes bursts out in so unexpected a manner at the sight of a person who laughs, and the frown which shows itself under similar circumstances, are also co-ordinated external revelations that follow upon an incident excitation carried into the *sensorium*, and reverberated towards the organs whose business it is to carry it off externally.

More than this—and this also is a phenomenon known to us all—in certain circumstances our muscles, which are usually such faithful interpreters of our wills, escape from the regular stimulation of the conscious personality, and then, under the influence of powerful emotions, become subject to invincible excitations radiated from the *sensorium*, and act like treacherous servants, only in obedience to the instructions of an irregular power, and manifest, without our consent, the different states through which our inner sensibility is passing. It is by reason of this substitution that our gestures, our movements, our attitudes, our physiognomy become, without our knowledge, living expressions of the different states of our sensibility, and in a manner apparent phenomena by which the phase of erethism of certain regions of the *sensorium* is externally discharged. In these cases our muscles of expression are grouped and harmonized in a co-ordinated manner, so automatically and so unconsciously that we see, for instance, those of the iris dilate and contract alternately, and express by their play, as automatic as unconscious, the different modes of sensibility of the retina which it is their business to protect.

We may say, then, in a general manner, that none of the peripheral excitations that arrive at the *sensorium* in the form of a vibratory impression, of a living force in activity, remain there stationary, stored up in one place. They develop there a series of secondary reactions, of energies regularly co-ordinated, which are incessantly distributed in the direction of the apparatuses of organic life, and represent the continuity of the primary movement, and, as it were, the modes of excretion of the living forces implanted in the organism, which here and there effect their physiological discharge.

Extrinsic Manifestations of Cerebral Processes. Genesis of the Will.—The processes of cerebral activity which reveal themselves externally, and make their exit from the organism in the form of *voluntary conscious manifestations*, must be considered successively in the two principal phases of their evolution:

1. In their period of incubation, when the process of the will is still only constituted by a purely physical impression;

2. In their second period of extrinsic manifestation, when they take form, reveal themselves in an apparent manner, and lay the purely motor regions of the nervous system under contribution.

1. In its preparatory phase of incubation, the process of the will is nothing but the riper and more advanced

ultimate period of an anterior operation of the judgment, constituted as we have already explained.

The human personality is seized upon by the arrival of the excitation emanating from the external world. It enters into participation and becomes associated with this; and from this intricate connection results a true intra-cerebral automatic radiation, which produces the apparition of a series of agglomerated secondary ideas. But the matter does not stop here; this inner personality having been thus seized upon, its sensibility having been touched in any manner whatever, has reacted by virtue of the vital forces that vibrate in it in a latent condition— it has been affected in the direction of its most profound affinities, and necessarily this reactionary period betrays itself by an unconscious desire for such or such a definite object, and by a repulsion from such or such another.

Desire, attraction, aversion, repulsion, are therefore new conditions of the *sensorium* which necessarily result in the natural course of things, and which thus become the primordial elements destined to constitute a process of voluntary activity.

2. The psychic operation which is to be resolved into an act of will is, then, in itself only the second bar of a movement already begun. It is only the regular expression of the human personality, seized on, and impressed by an old or recent excitation from the external world, and carrying back to the external world the different states of its sensibility in emotion, in the form of motor manifestations.

Hence, as a natural consequence, we come to the conclusion that the act of voluntary motion which is developed in the psychic regions, is nothing but a subordinate fact, a secondary phenomenon, the direct resultant of the shock of the sensibility in emotion and the spontaneous reaction of the *sensorium*. Motor power is then, physiologically, nothing but sensibility transformed. The voluntary excitation comes to life in that subtle process in which the impressed human personality is aroused. From this reaction of the sensibility it emerges as a natural consequence, like a vital force in evolution; it is like an excito-motor process radiating from the sensitive regions of the spinal axis towards the anterior regions, which progresses *motu proprio*, develops, amplifies, perfects itself infallibly through the whole length of its journey, and expands in its last period into co-ordinated motor manifestations, the faithful dependents of the sensitive excitations that have given it birth.

CHAPTER II.

TRUE PERIOD OF EMISSION OF THE PROCESSES WHICH PRODUCE VOLUNTARY MOTION. SPONTANEOUS REACTION OF THE SENSORIUM. MOTIVED RESOLUTION.

LET us now see how the different periods of voluntary activity are connected one with another, and how the physiological operation pursues its course.

The process of external emission of the emotivity of the *sensorium* manifests itself externally, sometimes in a rapid and instantaneous manner, sometimes slowly, progressively, and after a greater or less period of time; this extrinsic revelation taking place either in the oral or graphic form, or in the shape of gestures more or less expressive, and varied attitudes.

In the first case, when the voluntary motor phenomenon is an immediate translation of external impressions, the human personality, aroused and vibrating, rapidly responds to the impressions that affect it. It outwardly expresses itself directly, now in the form of connected articulate sounds, which are appropriate answers to the interrogations that excite it, now in current conversations, in injunctions of all kinds, prolonged discourses, in writings, expressive movements, etc., etc. It expends the stores of emotivity that are vibrating within it, and thus reflects the various sensitive currents that have set it vibrating.

Sensibility, therefore, underlies every motor act of the organism; and when we immediately answer to demands, when we let ourselves act upon the natural impulses of our sensibility, and, as it is called, do things on the spur of the moment, it is our personality that expands spontaneously, without artifice or premeditation. It reacts with its native and even frank characteristics, as though we had to do with physiological phenomena in natural evolution; for in these circumstances our words express our sentiments in an off-hand manner, and the compromises of meditation, and diplomatic reflection have not yet crossed our path to mask our natural spontaneity.

In a number of other cases the discharge does not take place in a rapid and immediate manner; there is, as it were, a *cold maceration* of the incident impression in the tissue of the *sensorium*, by which this impression is matured and modified by the mere action of the medium in which it remains.

When, in fact, we have to reflect, to mature a project, before coming to a resolution, the primitive idea, the first excitation, in arriving in the *sensorium* awakens a crowd of related reactions. It has been perceived in the form of sensorial vibrations, and these vibrations radiate to a distance into the different cell-territories. These latter, on being impressed, excite the automatic activity of those of the neighbourhood, and at the same time arouse related ideas and associated memories formerly registered; so that at the end of a period of sojourn in the *sensorium*, variable according to individual temperament, this primitive impression has proliferated and slowly produced effects that reverberate to a distance.

More than this, the ideas of others, in the form of oral counsels, written advice, and auditory and optic impressions interpreted by the intellect, have come to join in, to group themselves around the primary excitation, and add a new weight to the operation in process of development.

Those reflections which either proceed from ourselves, or are inspired by the surrounding medium, are then converted into agglomerated motives or thoughts, destined to influence the direction of the voluntary process and direct its route.

Things being thus disposed, a delicate phase occurs in the cerebral operation that is being accomplished. The motives being all confronted with one another, with their intrinsic and extrinsic characters, the shades which characterise them, their relative value, what route will the process take? Under what form will it reveal itself; and in what manner will the conscious personality pronounce itself? *

---

* This delicate moment of the operation, by virtue of which the *sensorium*, when seized upon, reacts spontaneously and carries outwards the different conditions of its impressed sensibility, does not occur in some individuals without certain difficulties.

There are a great many persons, indeed, whose hesitation is the dominant note of their character. At the moment of making a resolution they dare not decide, but turn about in a persistent indecision, and remain in suspense when action is necessary. In more pronounced cases, where this psychological condition is still more distinctly marked, we find individuals thus affected recounting all the anxieties that besiege them when they are on the point of coming to a decision. They hesitate, tormented by a series of uncertainties, and if they have to speak or take up their pen to affix a signature, or perform any spon-

On this point, the controversies of philosophers and metaphysicians, which have been taking place from time immemorial, have succeeded in arriving at but one thing—the expression in sonorous language of their ignorance, more or less complete, of the fundamental characters of psychical life. We must, indeed, penetrate into the inmost essence of the activity of cerebral life, into the complex phenomena in which it reveals itself, to arrive at a comprehension of the evolution of any voluntary act whatsoever, and the natural manner in which it expresses itself through the organism.

Little, indeed, as we may reflect upon the concatenation of the processes of cerebral activity, considered as we have here just done, we cannot help arriving at the conclusion, that the voluntary act is in itself nothing but the reaction of the sensibility thrown into agitation; that it is this that is latent in all voluntary manifestations; and that it is always the *sensorium* that, under forms the most dissimilar in appearance, reacts and outwardly betrays the inner impressions by which it is excited.

The sensibility is, therefore, always in agitation at the commencement of every voluntary act developed. It becomes erect, and excites the operations of judgment and reflexion. It is always present, always in vibration, and inspires our words, our acts, our

taneous action, they remain fixed, immovable, in a species of invincible apathy.

These different states, from the most simple to the most pronounced forms, are evidently only the effect of a partial or permanent weakening of the mental energies, through which the elements of the *sensorium*, in a torpid condition, are incapable of rising to the phase of erethism, of reacting, and of leading by their own vitality the process in evolution in the regular direction it should follow.

PROCESSES WHICH PRODUCE VOLUNTARY MOTION. 323

writings; and whatever be the power of the motives calculated to attract it away from its inner inclinations, it follows its preordained desires for what is suitable to it, what pleases it, and shrinks from what is repugnant to it. Every one, as we say, gives *his opinion*, every one judges according to the manner in which he is impressed, in which he *feels*; and sensibility, the seeking after what is pleasant to each of us, is, under the name of self-interest, to such an extent the true motive force of all human actions, that we may constantly declare that it is always this that directs them, like a powerful magnet, and inclines them in this way or that. All this takes place in so unconscious and certain a manner, that in dealing with a crime, or any guilty action, justice, *à priori*, ascribes responsibility to those who may have had an interest in committing it, by obtaining some profit from its perpetration.

On the other hand, since human sensibility is in itself one of the most mobile of things, and as regards this every one takes his pleasure as he finds it, it results that the manifestations of sensibility will vary infinitely according to individuals, and will sometimes assume paradoxical forms outside of the usual modes of common sensibility. But at bottom, although the sentiments of egotism and personal satisfaction may apparently be masked, the manifestations of the will will always demonstrate their derivation from the same origin. Everyone, as we have said, has his mode of feeling, and just as we see individuals experience satisfaction in certain enjoyments which they alone are capable of perceiving, so we find them manifesting these different states of their *sensorium* in eccentric and extravagant forms. Thus it is that the enthusiasms of generosity, self-abnegation, even self-sacrifice, are but too often only a disguised manifestation of egotism, a mode of *feeling, sui generis*, in which we exchange a physical advantage for an emotion of the moral kind.

From the moment, then, in which the personality becomes interested in the realization of such or such a desire, the moment in which, as we say, *a resolution has been taken* by it, this physiological condition expresses itself in a co-ordinated manner, according to processes which have been acquired by habit and commenced in infancy, and by which we have learnt to make our fellow-creatures comprehend by means of a special vocabulary the ideas which germinate in us, the desires that demand satisfaction, and our private aversions.

Henceforward the mental process has made one more step in the intricacies of the cortical substance. It opens up a new path, that of the motor regions proper. A living automatic pianoforte from this moment comes into play, and in various forms expresses the sensitive keys it is bound to interpret faithfully. It is the instrumental part of our organism that vibrates, and the process, tending more and more to emerge from the plexuses of the cortical substance, becomes concentrated within certain circumscribed limits, in certain psycho-motor regions, and hence, in the form of rapid intermittent stimulations, effects its discharge directly upon the different territories of the *corpora striata*.

Concatenation of Voluntary Motor Acts.—We have just seen how the voluntary stimulus, conceived in its primary phase of elaboration, in the substance of the plexuses

CONCATENATION OF VOLUNTARY MOTOR ACTS. 325

of the *sensorium*, as a condition of purely psychical vibration, was constituted by a series of multiple elements, all concurring in its genesis; how it became inevitably united with a previous phenomenon of sensibility in agitation; and how, like a living force in evolution, it tended more and more to emerge from the regions where it was conceived.

From this precise moment it leaves the purely psychomotor regions of the cortex, in the form of transient and rapid stimulations destined to be converted into articulate sounds, digital movements, or expressive gestures; and it proceeds, by help of the special white fibres (cortico-striate fibres), to different territories of the corpus-striatum, of which it thus excites the immediate activity. (See 5, 11, 16, Fig. 6, p. 61.)

Here, in this first stage of its outward course, it insensibly loses its original character of a purely psychical excitation, to incorporate itself more and more with the organism, to *materialize* itself, in a manner, and increase its dynamic power by the addition of a new nervous element, the cerebellar innervation, which, in the condition of a static force in permanent tension, is incessantly distributed in the plexuses of the *corpus striatum*.

Thus reinforced by this adventitious contingent of innervation which is engrafted into it, it continues its centrifugal course (see 7, 12, 19, Fig. 6, p. 61), and by means of the antero-lateral fibres of the axis (cerebral peduncles) it descends, in the form of an interrupted current, to excite the dynamic activity of the different motor nuclei of the spinal axis, which, like a series of apparatuses always ready to enter into action, only wait its arrival to develop their latent activity. From this moment, mixed up with the proper activity of the different spinal regions, it projects itself along the anterior roots and thus becomes, in its final phases of transformation, one of the multiple exciting causes of muscular contractility.

We see then, to sum up, from what precedes, that the processes which produce voluntary motion pass, in their evolution, through phases inverse to those of the processes of sensibility. While these latter, as they approach the central regions of the *sensorium*, are purified and made perfect, becoming more and more *intellectualized* by the metabolic action of the different nervous media through which they are propagated; the former, on the contrary, conceived as psychical vibrations at the moment of their genesis, amplify and are *materialized* more and more, as they descend from the superior regions. They become complicated by the addition of adventitious elements, which reinforce them as they progress (cerebellar and spinal innervation), and thus become, in the last term of their evolution, a true synthesis of agglomerated dynamic elements, which resume in themselves the vital forces of the system through which they have been developed—cerebral, cerebellar, and spinal activities.*

Conceived under this simple formula, the processes which produce voluntary motion begin by being a purely psychical excitation, and insensibly become, by the natural play of the organic machinery, a physical excitation. In thus becoming transformed in their successive evolu-

---

* See Luys, "Recherches sur le système nerveux cérébro-spinal," p. 434. (Iconographie photographique, p. 71.)

tion, they present the fascinating picture we constantly see presented to us in the working of steam-engines. We see, in fact, in this case, how a force, slight at its commencement, is capable of being transformed, and becoming by means of the series of apparatuses it sets at work, the occasion of a gigantic development of mechanical power.

In fact, at the moment when the engine begins to work, a very slight force, the mere intervention of the hand of the engine-driver who turns a handle and lets the steam rush against the upper surface of the piston, would suffice for this. This active force, once at liberty, immediately develops its strength, which is proportional to the surface over which it extends; the piston falls, its rod draws down the beam; the power is developed as the fly-wheel revolves, and the initial movement, so weak at its commencement, amplifies and increases continually, in proportion as the volume and power of the mechanical appliances placed at its disposal become more considerable and more powerful.

We see then, in conclusion, after an examination of all the details of cerebral physiology that we have successively passed in review, that the different processes of cerebral activity finally resolve themselves into a circular movement of absorption and restitution of forces. The external world, with all its incitements, enters into us by the channel of the senses, in the form of sensorial excitations; and the same external world, modified, and refracted by its intimate contact with the living tissues it has traversed, emerges from the organism, and is reflected outwards in the various manifestations of voluntary motor-power.

www.ingramcontent.com/pod-product-compliance
Lightning Source LLC
Chambersburg PA
CBHW031848220426
43663CB00006B/539